CAMBRIDGE LIBRARY COLLECTION

Books of enduring scholarly value

Botany and Horticulture

Until the nineteenth century, the investigation of natural phenomena, plants and animals was considered either the preserve of elite scholars or a pastime for the leisured upper classes. As increasing academic rigour and systematisation was brought to the study of 'natural history', its sub-disciplines were adopted into university curricula, and learned societies (such as the Royal Horticultural Society, founded in 1804) were established to support research in these areas. A related development was strong enthusiasm for exotic garden plants, which resulted in plant collecting expeditions to every corner of the globe, sometimes with tragic consequences. This series includes accounts of some of those expeditions, detailed reference works on the flora of different regions, and practical advice for amateur and professional gardeners.

Botanical Lectures

This reissue contains two works by the botanist Maria Elizabetha Jacson (1755–1829), a Cheshire clergyman's daughter. Her interest in science, and especially botany, may have been encouraged by a family connection with Erasmus Darwin, but it was not until she was in her forties that domestic circumstances drove her to professional writing. In 1797 she published *Botanical Dialogues, between Hortensia and her Four Children*, an introduction to the Linnaean system for use in schools. This technically rather demanding work was recast for adults in 1804 as *Botanical Lectures*: 'a complete elementary system, which may enable the student of whatever age to surmount those difficulties, which hitherto have too frequently impeded the perfect acquirement of this interesting science'. The more practical *Florist's Manual* (1816) was aimed at female gardeners, offering advice on garden design and the war against pests as well as notes on plants and cultivation.

Cambridge University Press has long been a pioneer in the reissuing of out-of-print titles from its own backlist, producing digital reprints of books that are still sought after by scholars and students but could not be reprinted economically using traditional technology. The Cambridge Library Collection extends this activity to a wider range of books which are still of importance to researchers and professionals, either for the source material they contain, or as landmarks in the history of their academic discipline.

Drawing from the world-renowned collections in the Cambridge University Library and other partner libraries, and guided by the advice of experts in each subject area, Cambridge University Press is using state-of-the-art scanning machines in its own Printing House to capture the content of each book selected for inclusion. The files are processed to give a consistently clear, crisp image, and the books finished to the high quality standard for which the Press is recognised around the world. The latest print-on-demand technology ensures that the books will remain available indefinitely, and that orders for single or multiple copies can quickly be supplied.

The Cambridge Library Collection brings back to life books of enduring scholarly value (including out-of-copyright works originally issued by other publishers) across a wide range of disciplines in the humanities and social sciences and in science and technology.

Botanical Lectures

And The Florist's Manual

MARIA ELIZABETHA JACSON

CAMBRIDGE
UNIVERSITY PRESS

University Printing House, Cambridge, CB2 8BS, United Kingdom

Published in the United States of America by Cambridge University Press, New York

Cambridge University Press is part of the University of Cambridge.
It furthers the University's mission by disseminating knowledge in the pursuit of
education, learning and research at the highest international levels of excellence.

www.cambridge.org
Information on this title: www.cambridge.org/9781108067027

© in this compilation Cambridge University Press 2014

These editions first published 1804 and 1816
This digitally printed version 2014

ISBN 978-1-108-06702-7 Paperback

Selected botanical reference works available in the
CAMBRIDGE LIBRARY COLLECTION

al-Shirazi, Noureddeen Mohammed Abdullah (compiler), translated by
Francis Gladwin: *Ulfáz Udwiyeh, or the Materia Medica* (1793)
[ISBN 9781108056090]

Arber, Agnes: *Herbals: Their Origin and Evolution* (1938)
[ISBN 9781108016711]

Arber, Agnes: *Monocotyledons* (1925) [ISBN 9781108013208]

Arber, Agnes: *The Gramineae* (1934) [ISBN 9781108017312]

Arber, Agnes: *Water Plants* (1920) [ISBN 9781108017329]

Bower, F.O.: *The Ferns (Filicales)* (3 vols., 1923–8) [ISBN 9781108013192]

Candolle, Augustin Pyramus de, and Sprengel, Kurt: *Elements of the Philosophy
of Plants* (1821) [ISBN 9781108037464]

Cheeseman, Thomas Frederick: *Manual of the New Zealand Flora*
(2 vols., 1906) [ISBN 9781108037525]

Cockayne, Leonard: *The Vegetation of New Zealand* (1928)
[ISBN 9781108032384]

Cunningham, Robert O.: *Notes on the Natural History of the Strait of Magellan
and West Coast of Patagonia* (1871) [ISBN 9781108041850]

Gwynne-Vaughan, Helen: *Fungi* (1922) [ISBN 9781108013215]

Henslow, John Stevens: *A Catalogue of British Plants Arranged According to
the Natural System* (1829) [ISBN 9781108061728]

Henslow, John Stevens: *A Dictionary of Botanical Terms* (1856)
[ISBN 9781108001311]

Henslow, John Stevens: *Flora of Suffolk* (1860) [ISBN 9781108055673]

Henslow, John Stevens: *The Principles of Descriptive and Physiological Botany*
(1835) [ISBN 9781108001861]

Hogg, Robert: *The British Pomology* (1851) [ISBN 9781108039444]

Hooker, Joseph Dalton, and Thomson, Thomas: *Flora Indica* (1855)
[ISBN 9781108037495]

Hooker, Joseph Dalton: *Handbook of the New Zealand Flora* (2 vols., 1864–7) [ISBN 9781108030410]

Hooker, William Jackson: *Icones Plantarum* (10 vols., 1837–54) [ISBN 9781108039314]

Hooker, William Jackson: *Kew Gardens* (1858) [ISBN 9781108065450]

Jussieu, Adrien de, edited by J.H. Wilson: *The Elements of Botany* (1849) [ISBN 9781108037310]

Lindley, John: *Flora Medica* (1838) [ISBN 9781108038454]

Müller, Ferdinand von, edited by William Woolls: *Plants of New South Wales* (1885) [ISBN 9781108021050]

Oliver, Daniel: *First Book of Indian Botany* (1869) [ISBN 9781108055628]

Pearson, H.H.W., edited by A.C. Seward: *Gnetales* (1929) [ISBN 9781108013987]

Perring, Franklyn Hugh et al.: *A Flora of Cambridgeshire* (1964) [ISBN 9781108002400]

Sachs, Julius, edited and translated by Alfred Bennett, assisted by W.T. Thiselton Dyer: *A Text-Book of Botany* (1875) [ISBN 9781108038324]

Seward, A.C.: *Fossil Plants* (4 vols., 1898–1919) [ISBN 9781108015998]

Tansley, A.G.: *Types of British Vegetation* (1911) [ISBN 9781108045063]

Traill, Catherine Parr Strickland, illustrated by Agnes FitzGibbon Chamberlin: *Studies of Plant Life in Canada* (1885) [ISBN 9781108033756]

Tristram, Henry Baker: *The Fauna and Flora of Palestine* (1884) [ISBN 9781108042048]

Vogel, Theodore, edited by William Jackson Hooker: *Niger Flora* (1849) [ISBN 9781108030380]

West, G.S.: *Algae* (1916) [ISBN 9781108013222]

Woods, Joseph: *The Tourist's Flora* (1850) [ISBN 9781108062466]

For a complete list of titles in the Cambridge Library Collection please visit:
www.cambridge.org/features/CambridgeLibraryCollection/books.htm

Plate 6. *Part I. P.109.*

Lycopérdon Fornicatum.

*Found growing in Mr. Rooke's
Kitchen Garden, near Mansfield Woodhouse,
September 1792.*

London, Published May 1. 1797. by J. Johnson, St. Paul's Church Yard.

BOTANICAL LECTURES.

BY A LADY.

ALTERED FROM

"BOTANICAL DIALOGUES FOR THE USE OF SCHOOLS,"

AND

DAPTED TO THE USE OF PERSONS OF ALL AGES,

BY THE SAME AUTHOR.

———————

LONDON:

PRINTED FOR J. JOHNSON, ST. PAUL'S CHURCHYARD,

BY T. BENSLEY, BOLT COURT, FLEET STREET.

———

1804.

ADVERTISEMENT.

From the favourable reception given to my Botanical Dialogues for the Ufe of Schools, I have been induced to fuppofe, that the Work might be of more extenfive utility, if divefted of thofe parts peculiarly intended for the purpofes of education, and altered to a form equally adapted to the ufe of grown perfons as to children. I have, therefore, endeavoured to compofe a complete elementary fyftem, which may enable the ftudent, of whatever age, to furmount thofe difficulties, which hitherto have too frequently impeded the perfect acquirement of this interefting fcience; and I flatter myfelf that the following Work, in *Botanical*

7 *Lectures,*

Lectures, will be found an eafy introduction to the ufe of the Tranflated Syftem of Vegetables, the only Englifh work from which the pupil can become a Linnean, or univerfal Botanift.

M. E. J.

Oct. 1, 1803.

ANALYSIS OF THE FIRST PART

OF THE

BOTANICAL LECTURES.

===

LECTURE I.

The Seven Parts of Fructification explained.

Pᴀɢᴇ 1, Explanation of Linneus's fyftem. 2, The Linnean terms ought to be made ufe of; knowledge of Latin of no great ufe to botanical pupils. 3, Botany has a language pe-culiar to itfelf. Term Fructification explained; all parts of it not effential to the product of perfect feed; feven parts of Fructification; Calyx, feven different kinds; thefe explained. 4, Fool's Parfley diftinguifhed from all other known umbelled plants. 5, Male Bloom of Willow, called Yellow Goflings by children; beauty of different Catkins. Spathe explained. 6, Calyptre, the Calyx of Moffes beautifully fhown in Mr. Cur-tis's London Flora. Volve the Calyx of Fungufes. 7, Corol explained; it's leaves called petals. Marks by which the different kinds of Corol are diftinguifhed. Polyanthos a one-petalled Corol. Method of knowing a one-petalled Corol from a Corol of many Petals. 8, Génera of Plants diftin-guifhed by the form and pofition of their Petals. Seven dif-ferent formed Corols. 9, Nectary, the name given by Linneus to the honey-bearing part of the Corol; not always

a   a part

a part of the Corol. Stamen, a moſt eſſential part of
Fructification, conſiſts of three parts. 10, Anther, wrongly
called the Seed, explained. Nature has provided for the
ſecurity of the Duſt. On it's preſervation depends the continuation of the ſpecies. Wet injurious to the Anther-duſt.
Suppoſition of the Anther-duſt being preſerved from injury
by a waxy ſubſtance ſurrounding it, erroneous. Mr. John
Hunter's experiments prove that the Anther duſt is not Wax.
Collected by Bees for Food to the Bee-maggots. 11, Piſtil
of equal importance with the Stamen, conſiſts of three
parts. Anther and Stigma, in botanical language, conſtitute
a Flower; eſſential to the production of Fruit—late inveſti-
gations ſeem to make the Nectary an eſſential part of Fructification; Honey contained in it intended for the nouriſhment
of the Anthers and Stigmas. 12, The Nectary had not even
a name before the time of Linneus. Eight different
kinds of Seed-veſſel. Sudden manner of the Seed-veſſel of
Balſam burſting. 13, Two kinds of Silique explained.
Diſtinction between the Silique kind of Seed-veſſel and the
Legume. Legume the part eaten of the Papilionaceous, or
Pea-bloom Flowers. Follicle and Drupe explained. 14, Exceptions to the definition of a Drupe. Defects of the ſyſtem
of Linneus few. 15, Strobile, the Strobiles of Larch, beautiful, commonly called Fir-apples. Seed, definition of it.
Seed conſiſts of three parts. 16, Young plants ſupported in
a ſimilar manner to young animals. By ſoaking an almond
kernel in Water, the three parts of the Seed may be well ſeen.
17, Young Plants periſh, if their Seed-lobes or Cotylédons
are deſtroyed. Corn dug out of the ground by Wood Pigeons.
Care taken by Nature in the protection and diſperſion of
Seeds. Muſlin made from Cotton. Cotton the ſoft Cradle of
Seeds, as Silk is that of Inſects. Aril, that part which ſur-
rounds the Seed within it's Veſſel, may be ſeen in Fraxinella
and Wood ſorrel. 18, Diſperſion of Seeds. 19, Beauty of
the Seed of Feather-graſs: moſt curious of the Flying Seeds,
Tillándſia, a paraſite plant: manner in which it's ſeeds are
conveyed

conveyed fimilar to the migration of Spiders. 20. Beautiful
lines in the Botanic Garden, on the migration of the feeds of
aquatic plants, and of thofe which grow on the banks of rivers.
Much knowledge to be gained from the Botanic Garden.
Birds the means of diffeminating fome kind of feeds.
21, Mountain Afh on an Apple-tree. Doubtful in what
manner Trees growing upon others receive Nourifhment.
Seeds of Feather-grafs, Geránium, and Oat, diflodged from
their Receptacles by the twifting of their Awns. 22, Two
kinds of Receptacle, Proper Receptacle, and Common Receptacle; Common Receptacle belongs peculiarly to the Compound Flowers. Examples of a Common Receptacle, made
ufe of by Linneus to difcriminate the Génera of the Clafs Syn-
genefia, United Anthers.

LECTURE II.

A Flower diffected. Fulcra and Inflorefcence explained.

PAGE 27, Specimens of the different kinds of Calyx.
28, Peculiarity in the Wheel-form Corol of Verónica. The
Genus diftinguifhed by that peculiarity. Hollow protuberance at the bafe of the Petals of Ranúnculus, the
Nectary. The effential Mark of that Genus, conftant, even
in the double Flowers. 29, Flowers more eafily to be diftinguifhed from each other than is fuppofed by thofe who have
not attentively ftudied their different parts. 30, Specimens
of different formed Corols. Corol not always coloured; no
obvious rule by which the Calyx and Corol may be diftinguifhed; rule given by Linneus for diftinguifhing them;
31, his tranfgreffion of that rule; advifable to follow the
terms of Linneus. Cover of the Crown Imperial, a Corol.
32, Quantity of Honey in the Nectaries of Crown Imperial
nicely adapted to the Cavities which contain it. Form of

the

the particles of Anther-duft perfect and regular. Moiflure
on the fummit of the Stigma fits it to receive the Anther-duft.
Germe, the term for the immature Seed-veffel; Pericarp, for
that which is mature. 33, Only fix parts of Fructification
in the Flower of Crown Imperial. Bract, a part which may
be miftaken for a Calyx. Rule for diftinguifhing the Bract
from the Calyx. Green Tuft at the Top of the Stem of the
Crown Imperial formed from Bracts; the Bract a part of
great ufe in marking the Species of Plants. 34, Bract one of
the Fulcra of Plants. Poppy and Tulip fhow the Stigma and
Germe without a Style. Great increafe of Plants from Seed.
Beauty of Seeds; great variety in the fize, fhape, and furface
of Seeds. 35, Explanation of the term Fulcra. Seven kinds
of Fulcra. Diftinction betwixt the terms Peduncle and Petiole. 36, Stipule explained. Stipules of Plants ought to be
attended to, as they frequently mark one fpecies from another.
Infant Leaves of Tulip-tree perfect in all their parts. Stipules
of Plane-tree add to it's beauty in fpring. Weak Plants generally furnifhed with Tendrils. 37, All climbing Plants injurious to the Trees which fupport them. Dodder feems intended
by Nature to draw nourifhment from other vegetables. 38, By
it's growth, ftrangles the Plant by which it was foftered. Few
inftances of this in the Vegetable Kingdom. The injury
fuftained by the fupporting Plant generally fmall. Orobánche
a parafitical Plant from choice. 39, Pubefcence might more
properly be called a Defence than a Support. Young Leaves
and Stems commonly protected by a downy Covering.
40, Arms of Plants, Thorns and Prickles their Defence againft
Animals. Hollies in Needwood Foreft, armed only on their
lower branches, afford food to the Deer in fevere winters.
41, Curious mechanifm of the Sting of a Nettle. Many
curious contrivances of Nature for the defence of Plants.
Venus's Fly-trap particularly curious. 42, Sun-dew refembles
it; eafily found on heaths by the red colour of it's leaves.
43, Scape, a particular kind of Flower-ftalk. Inflorefcence,
the term for the different modes by which Flowers are joined

to

to their Peduncles. Seven different kinds of Inflorefcence.
44, One-ranked and two-ranked Spike explained. 45, Diftinction betwixt a Corymbe and Umbel. Thyrfe and Raceme
explained. 46, Wherein the Raceme and Corymbe differ.
Panicle explained. Modes of Flowering not comprifed under
the term Inflorefcence. 47, Linneus's definition of the term
Rachis. 48, Method of impreffing what is taught, upon the
memory. The art of making learning agreeable; the more
deeply the ftudy of Botany is entered into, the more pleafing
it will be found.

<hr>

L E C T U R E   III.

The firft Eighteen Claffes with their Orders explained.

PAGE 51, Explanation of the term Clafs: may be compared
to a Dictionary. Characteriftic mark of a Clafs arbitrary.
On the number and fituation of the Stamens the Claffes of
Linneus are founded; 52, what conftitutes a natural Clafs;
moft of the Claffes of the Linnean fyftem artificial; their being fo of little confequence; the great advantages of his fyftem.
Labours of many ingenious Botanifts of little ufe from want
of arrangement. 53, Much ufeful knowledge of the ancients
loft to the world from their ignorance of the fcience of Botany. Dr. Grew's book very informing. His opinion of the
ufe of the parts of Fructification agrees with that of Linneus.
Linneus's works beft calculated to teach the fcience of Botany.
54, Linneus divided the vegetable kingdom into twenty-four
Claffes. Character of the firft ten Claffes. Names of the Numerical Claffes taken from the number of Stamens or Males.
55, Ufeful to be acquainted with the fcientific terms of
Botany. The tranflated Syftem of Vegetables found difficult
from not being properly ftudied. 56, Ten firft Claffes diftinguifhed by their number only. Eleven Stamens not found

fufficiently conftant to form a Clafs. 57, Titles of the laft three Numerical Claffes would miflead if they were not explained. Linneus aware of the defe $\mathcal{E}$ in the titles. Diftinctions of the clafs Icofandria and Polyandria neceffary to be attended to. 58, Fruits belonging to Icofandria have their Calyx remaining when ripe, like a little crown on their fummit. 59, Great number of Stamens in the clafs Polyandria. Explanation of the Orders or firft Subdivifions of the Numerical Claffes founded on the number of Piftils. 60, A Flower cannot belong to any of the firft thirteen Claffes, unlefs it contain both Stamens and Piftils within the fame cover. Effential Charaĉer of the Eleventh Clafs. Effential Charaĉer of the Twelfth and Thirteenth Claffes. 61, Charaĉer of the clafs Two powers; contains two Orders. Diftinguifhed by their feeds being enclofed by a Veffel or not enclofed. Flowers of the different Orders not fimilar in their appearance. 62, Clafs Tetradynamia, or Four-powers, explained, a really natural Clafs; no exception to this, except the Genus Cléome. Divided into two orders from the form of the Seed-veffels. 63, A good deal of variety in the form of the Silicle. Two divifions. of the Silicle Order. Seedveffel of Lady Smock, a Silique. Clafs Monadelphia, or One-brotherhood, explained; beauty of the pofition of the Stamens and Piftils of this clafs; peculiar ftruĉure of the Anthers. 64, Stamens and Piftils of China Rofe particularly beautiful. Orders of the clafs Monadelphia founded on the number of Stamens in each Flower. Clafs Diadelphia, or Two-brotherhoods, perfeĉly natural. Peculiar Struĉure of the Flowers belonging to it. 65, It's claffic charaĉer difficult to be traced. Genus Sophóra feparated from the Twobrotherhood Clafs, from it's Filaments not being united; the Orders founded on the number, only, of Stamens; each part of the Corol diftinguifhed by different names. 66, Shape, &c, of thefe parts of ufe in marking the Génera, particularly the Calyx. Legume belongs to the Diadelphia Clafs of Plants. Diftinĉion betwixt the Legume and Silique Seed-veffels.

Clafs

Clafs Polyadelphia, or Many-brotherhoods. St. John's Wort, a good Specimen of this Clafs. The Orders of Polyadelphia depend on the number of Anthers in each Flower. 67, Anthers and Stigmas the effential parts of the Stamens and Piftils.

———

LECTURE IV.

Examination of Flowers of different Claffes. Claffes 19, 20, 21, *and* 22, *explained.*

PAGE 68, Rules for the inveftigation of Flowers. Hippúris Vulgáris, remarkable for the fimplicity of it's ftructure. 69, Specimens of flowers belonging to different claffes. 70, Deep divifions of the Stigma of Crocus make the order to which it fhould be referred doubtful to young Botanifts; it's Fructification cannot be accurately examined without taking the root out of the earth. Stamens of Plantágo (Plantain), curioufly folded within the Corol. 71, Difficulty of inveftigating the Umbel-bearing Plants. The terminating Flower of the Umbel determines the clafs to which the Flower belongs. Number of Stamens often varies in Flowers of the clafs Pentandria. More Flowers than one of the fame Plant fhould be examined. 72, Umbelled Plants not proper fubjects to begin with. Large Flowers of fimple conftruction fhould be firft examined. Specimens of Plants belonging to different claffes. Clafs Enneándria (Nine Stamens) contains only fix Génera. Only one Britifh fpecies of this clafs. 73, Lychnis Dioica puzzling to young Botanifts, being placed in the clafs Ten Stamens; a defect in the fyftem. Obviated by being noted for it's want of Piftils. Lythrum fubject to vary in it's number of Stamens; neceffity of examining many Flowers of the fame genus. 74, Marks of the Twelfth and Thirteenth Claffes. Specimens of different Claffes and Orders, not depending on

a 4   the

the number of Stamens. 75, Seed-veffel of Drába (Whitlow Grafs) a Silicle. Seed-veffel cf Héfperis (Purple Rocket) a Silique. Many Plants of the clafs Tetradynamia, Four-powers, eaten, fome without cookery; variety of eatable plants from the genus Bráffica. Change produced in plants by the art of gardening an amufing part of the ftudy of Botany. Specimens of the clafs Monadelphia, One brother-hood. Stamens firmly united at the bafe. 76, Syftematic charaĉter of the clafs Diadelphia, Two-brotherhoods, fhown in Pea. 77, Curious circumftance refpeĉting the Piftil of Common Broom. 78, Specimen of the clafs Polyadelphia (Many brotherhoods). Stamens of Hypericum (St John's Wort) beautiful; the only Britifh genus of the Polyadelphia clafs. The genus Cítrus, comprifes Orange, Lemon, and Citron. Different appearance of their Stamens to thofe of Hypéricum. 84, Explanation of the clafs Syngenéfia, or United Anthers. Elafticity of the Filaments in the Flowers of this clafs; 79, confifts of the Compound Flowers; natural, if a few Génera be excepted; this exception a fault in the fyftem. What conftitutes a Compound Flower. 80, Genèric charaĉter founded, in part, on the variety in the form of the Corol. The firft four Orders founded on the Stamen-bearing and Piftil bearing Florets. Mark of the Fifth Order. Sixth marked by the Corols being fimple. 81, Perhaps, from th's circumftance, ought to have been feparated from the Clafs. Placed in it by their Anthers being united. Linneus does not pretend to make his Claffes natural. Gratitude due to Linneus from all Botanifts due alfo to his predeceffors. Tournefort's fyftem ingenious. Orders of the clafs United Anthers cannot be retained by the memory without examining flowers belong-ing to each. 82, Scabious has the appearance of a Compound Flower; belongs to the clafs Tetrándria, Four-ftamens. Marked diftinĉtions between them. 83, Scabious, a fpecimen of an Aggregate Flower. Specimens of the orders of the clafs United Anthers fhould be ftudied according to their orders. Florets of the Fourth Order having Stamens and Piftils not the

only

only circumſtance to be attended to. Having ſeeds or not, the
eſſential charaĉter of the Fourth Order. Charaĉters of the
Orders. 84, Globe Thiſtle (Echinops). Difference in the
Stigmas of Violet and Panſie. Jaſíone cannot belong to an
order of Simple Flowers; it's Anthers united only at the baſe;
does not agree exaĉtly with the charaĉters of the Compound
Flowers, or thoſe of the Aggregate. 85, Curious circumſtance of the Calyx of Compound Flowers. 91, Elegant
forms of thoſe Compound Flowers which have their ſeeds
furniſhed with a Pappus. 86, Extraordinary ſtruĉture of the
Flowers of the claſs Gynándria; it's eſſential charaĉter. The
Piſtil muſt be firſt attended to. Contains nine orders, founded
on the number of Stamens. Firſt Order natural. Struĉture
of the Fruĉtification explained. 87, Reſemblance of the
Flowers of the Firſt Order to inſeĉts; fanciful names given
them. Ophrys Genus contains ſeveral ſpecies reſembling
inſeĉts. Neĉtary, the principal feature in their different
forms. Bee Ophrys. Cypripedium has it's name from it's reſemblance to a ſlipper. 88, Struĉture of the parts of Fruĉtification of Arum differs from that of all other known plants.
Opinion of the younger Linneus reſpeĉting it. Plants growing commonly on the hedge-banks ſhould be well underſtood.
89, Explanation of the claſs Monoecia, or One-houſe. Orders
founded on the number, union, and ſituation, of the Stamens.
Eleven orders of the claſs One-houſe. Names by which they
are diſtinguiſhed. 90, Eſſential charaĉter of the firſt twenty
Claſſes. Deſcription of the true Nutmeg, Myríſtica, firſt
given by Dr. Thunberg. Claſs Dioecia, Two-houſes, explained.
Flowers of Valliſnéria thought to be a ſtrong argument for the
ſenſation of plants. Hemp, Cánnabis, and Willow, Salix, belong to the claſs Two-houſes; contains fifteen orders founded
on the number, union, and ſituation, of the Stamens. Contra-
diĉtions in the ſyſtem of Linneus; 91, a removal of it's defeĉts
may be expeĉted from the liberal ſpirit of the preſent age. Miſle-
toe, Viſcum Album, a paraſitical plant. Can be propagated only
by one method; curious manner of the ſeed germinating.

LECTURE

LECTURE V.

Claſs Polygamia explained.   Claſs Cryptogamia explained.

PAGE 95, Explanation of the claſs Polygamia. Many plants of this claſs diſperſed into the claſſes Mcnoecia and Dioecia. 96, Difficulty of aſcertaining by what manner the Anther duſt of the Fig, Ficus Carica, was conveyed to the Stigmas of the Piſtils. Fruit of the Fig a Receptacle encloſing the Stamens and Piſtils. Fertilization of it's ſeed ſuppoſed to be effected by the intervention of a Gnat. Proceſs performed by it, termed Caprification; object of much attention to the inhabitants of thoſe countries in which Figs make an article of trade. 97, Account of Caprification given by Mr. Milne. Objections againſt the neceſſity of Caprification. Receptacle of Figs gapes at top when the Stamens are mature, analogous, in this, to water plants. Air, an element apparently neceſſary to the proceſs of fertilizing ſeeds. Caprification eſteemed by many authors a ſtrong argument for the ſyſtem of Linneus. Firſt doubted of by the author of the Botanical Garden: his conjecture concerning it. 98, Apples wounded by worms ripen ſooner than others which are not ſo. Fig trees of Malta bear two crops in the ſame ſeaſon; laſt crop ripened by Caprification. Figs of Provence and Paris ripen ſooner by being wounded with a ſtraw. Probable that the ſecond crop of Figs in Malta ripens from being pierced. 99. Fig-trees cultivated in England produce two crops; latter crop pulled off by gardeners. Crop obtained by Caprification in Malta ſcanty, and not of good quality. The flowers of Fig to be looked for within the part which is eaten as fruit. Inſide of a Fig beautiful. Antherduſt may be ſeen in the Figs cultivated in England, if opened when they gape at top. 100, Claſs Cryptogamia explained; conſiſts of four orders. The ſyſtem of Linneus may have retarded a more diſtinct knowledge of this claſs. Definition of
Ferns.

Ferns. 101, Leaf of Fern termed by Linneus a Frond. Curious mechanifm of the feed of Ferns. Sago Powder made from the pith of a fpecies of Fern. Vegetable Lamb, a fpecies of Fern; marvellous ftories from want of proper inveftigation. 102, Glove and Stocking-tree in Caffraria. Confufion arifing from too great credulity; facts fhould be reafoned upon before they are affented to. Root of Common Fern (Ptéris Aquilina) ufed for bread in New Zealand. Bread made from a fpecies of Fern in the Canary Iflands. Second Order of Cryptogamia contains the Moffes, Mufci; circumftances from which the Génera are marked. 103, Their feeds have no Cotylédons. Linneus doubted whether what he termed the Anthers were really fo. Dillenius the firft who attempted the arrangement of the Moffes. Many curious circumftances belonging to the tribe of Moffes; recover their verdure on being moiftened, after having been long dried. Fructification of the Algæ, Flags, too obfcure to admit of precife arrangement; two divifions of them. Terreftrial and Aquatic, their Génera diftinguifhed by the outer ftructure. Many curious and ufeful Vegetables among the Algæ. Lichen Rangiferinus, or Rein Deer Lichen, it's ufe to the inhabitants of northern climates. 104, Different fpecies of Lichen ufed in dying. A fpecies of Ulva ufed for food by the Japanefe; fome kinds ufed for pickles in England, 105, Curious ftructure of fome of the Aquatic Algæ. Conférva Ægagrópila, Vagabunda, and Fúcus Natans, itinerant vegetables. Byffus Flos-aquæ, floats on the fea all day and finks at night. Conférva Polymórpha, lines upon it in the Botanic Garden; 106, grows on the Britifh fhores. Laft Order of Cryptogamia confifts of the Fungufes, Fungi, divided by Linneus after the method of Dillenius. Method of Dillenius explained. Fungus tribe divided into ten Génera. Fungufes produced from feed; 107, their fpecies conftant; renewed by uniform laws; little known of this part of the vegetable creation. Much attended to in thefe times. Mr. Curtis's inveftigations valuable on this fubject. Late difcoveries of the production of animals may lead, by analogy, to

the

the knowledge of the reproduction of vegetables. Curious facts of the Polypi genus; 108, Experiments of Monfieur Trembley. Hydra, the Linnean name of the Polypi genus. Reproduction of Plants from Strings and Suckers, fimilar to the increafe of Polypi. Information to be gained of the clafs Cryp ogamia very fmall. Parts of Fructification not only to be confidered. Experiments founded on analogy may lead to important difcoveries. Small progr fs made from thofe which prefuppofed a Fructification. 109, Beauty of the Cryptogamia Plants in winter. Extraordinary ftructure of Lycoperdon Fornicatum, Appendix of Linneus. Plants contained in it arranged under the general head of Palms. Singular ftructure of thefe plants. Their leaves refemble thofe of Ferns. Termed Fronds; their Fructification produced on a Spadix. 110, Terms Spathe and Spadix originally applied to Palms only, now ufed for other plants, whofe flowers are protruded from a Sheath. Cocoa nut, Cocos Nucifera, and Date-tree, Phœnix Dactylifera, Palms. Anther-duft of Date-tree, and Piftácia, faid to retain its virtues more than a year. Great height of Cory'pha Umbraculifera. Erroneoufly named Cabbage tree. True Cabbage Palm, *Aréca Oleracea*. 111, Ufed by the inhabitants of the Weft Indies as a rarity; fent pickled to Europe as fuch. Cutting away the Cabbage fhoot deftroys the tree. 112, Cabbage obtained from moft of the Palms. Breadfruit tree, Artocárpus Communis, of Forfter. 113, Has born fruit in Jamaica. Difappointment of Dr. Thunberg, in his attempt twenty years ago, to bring Breadfruit trees from Ceylon into Europe. The fruit made ufe of in cookery by the rich inhabitants of Ceylon. Fifteen different difhes prepared from it in Ceylon. The fruit of extenfive benefit to the poor. Make ufe of it as the poor of England do of potatoes. Ufed by the natives of Otaheite in a fimple manner. 114, Two kinds found in Ceylon; the leaft fort without feeds, the larger produces great numbers of feeds; fize of the feeds. Several varieties of the Artocárpus in the South Sea ifles, all without feeds; this deficiency attributed by Mr.

Forfter

Forſter to the effects of cultivation. The Breadfruit-tree of Ceylon ſuppoſed to be of the ſame genus with that of Otaheite. Seeds of the Breadfruit-tree of great value; eaten by the rich; prepared in different ways; eaten plain roaſted by the poor; ſimple manner in which the Breadfruit is uſed by the poor inhabitants of Ceylon. 115, The trees flouriſh whole centuries; bear fruit on their Stems. The fruit uſed for food in three different ſtates of maturity; when quite ripe eaten in it's freſh ſtate. Plaintain tree, Muſa Paradiſiaca, and Banana, Muſa Sapientum, called Bread-trees in the Weſt Indies. Cultivated in Jamaica for the uſe of the negroes; found in the South-Sea iſles. Banana loſes it's ſeeds by cultivation. 116, Leaves of Banana made uſe of for ſhade in warm climates. Cocoa-nut-tree deſerves a place in the firſt rank amongſt the vegetables which are uſeful to mankind. Leaves of Boraſſus Flabelliformis, and Licuála Spinoſa, uſed by the inhabitants of Ceylon, in the ſtate in which they grow, for writing upon. Ingenious method of writing upon them; books made of them. Leaves of Licuála uſed for umbrellas; ſix perſons may be ſheltered by one of theſe leaves. 117, Natural Orders, attempted by Linneus, placed at the end of the Génera Plantárum. Natural method attempted by many Botaniſts not without ſucceſs. Merit of artificial ſyſtems generally allowed. Opinion of Linneus concerning natural ſyſtems. Fifty-eight natural Orders of Linneus. 118, Theſe Orders well explained in Mr. Milne's Botanical Dictionary. Artificial ſyſtem muſt firſt be learnt. Order in which a young Botaniſt ought to proceed.

ANALYSIS

OF THE

SECOND PART.

───────────

LECTURE I.

Génera of Plants.

Page 119, Génera of Plants, the Third Divifion of the Syftem; the term Genus explained. 120, Botanical Alphabet of Linneus. Receptacle of the Fructification explained. 121, Receptacle of the Flower and of the Fruit explained; made ufe of in the Génera Plantarum only when it forms a Character of the Genus. Botanical Alphabet, or 26 Marks, taken from the parts of Fructification. Effential Characters. 122, Language of the tranflated Syftem of Vegetables excellent. Canna and Hippúris proper plants for examination; manner of referring them to their Genus. 123, Terms *above* and *beneath* explained. Manner of referring a plant to it's Genus continued. Departure from the general rule of the Syftem always noted by Linneus. 124, Genéric Characters of Canna. Hippúris eafy to be referred to it's Genus; merit of the Linnean method. 125, Honeyfuckle referred to it's Genus. Diftinguifhed from Triófteum by the form of the Stigma. 126, Specific diftinctions of Lonicéra; remarks on the fpecies in the Génera Plantarum. 127, Species Plantarum not tranflated. Wonderful ingenuity of the Syftem of Vegetables. Iris a perplexing flower to the young Botaniſt; fhould be referred to

it's

it's Genus in the manner recommended in Canna and Lonicéra.
Effential Charaĉter of Iris; defcription of it. 128, Different
kinds of Neĉaries found in Iris. Variety in the form of the
Capfules. Diftinĉive marks of Génera noted. Marks from
colour, fmell, and tafte, of Plants rejeĉed by Linneus as being
too inconftant; 129, dimenfions and place of growth not admitted, from their uncertainty; not without their ufe. Nectaries of importánce in the Genéric Charaĉer. Orders of the
clafs Monadélphia. Marks of difcrimination of the Génera.
130, Arrangement of the genus Geránium by Linneus.
L'Heritier's improvement of the Geránium family; his arrangement explained. Doubts refpeĉing the claffical mark of
Geránium. 131, Four fpecies of Britifh Geránium removed
to the genus Eródium. Dr. Smith's improvement of the fpecific diftinĉions of Geránium. His Englifh Botany. Marks
of the Orders of the clafs Syngenéfia, United Anthers.
132, Artichoke referred to it's Genus. Beauty of the Pappus.
Dandelion referred to it's Genus. 133, Effential Charaĉers
of Dandelion. Pappus of Dandelion. Diftinĉion of the terms
plumy and *hairy*. Dandelion deficient in one of the effential
charaĉers of it's Genus. Deviation of different fpecies of
Leóntodon from the marks of their family. 134, Plumy
Pappus in Tragopogon. Hairy Pappus of Artichoke. Diftinction of the Pappus of great importance. Method of difcerning
the diftinĉion. Scopoli's opinion of the deficiency in the
plumy Pappus in Dandelion. Obvious charaĉeriftic diftinctions of the Compound Flowers. 135, Mr. Curtis's difcoveries
of different marks in thefe flowers. Smaller flowers of the
clafs Syngenéfia lefs eafy of inveftigation than the larger
kinds. Umbelled plants. Mode in which they fhould be inveftigated. 136, Charaĉers of the fubdivifions of the order
Digynia of the clafs Pentándria. Explanation of various terms
ufed in the charaĉers of the Umbelled Tribe. Explanation of
the fame terms applied to the Compound Flowers. 137, Terms
uniform and *not uniform*, explained. Form of the Seeds of confequence in the fpecific charaĉer of Umbelled Flowers. The
Radiate

Radiate Flower fhown in Scandix Peften, Shepherd's Needle.
138, Difk and Ray explained. Difference in the fpecies of
Gentiána. Centaury removed to the genus Chirónia. The
accurate obfervations of Mr. Curtis. His candid criticifms of
the works of Linneus.

L E C T U R E   II.

Neftaries of Plants.

PAGE 140, Extraordinary appearance of the Stamens of
Houfeleek explained by Mr. Curtis. 141, Advantage of examining flowers in different ftates of maturity. Diftinétion
betwixt Sempervívum and Sedum. Genus Euphórbia accurately defcribed by Mr. Curtis. Linnean charaéters of Euphórbia defeétive. 142, Inveftigation of Euphórbia, on the Linnean
principles, extremely difficult; a diftinét idea may be attained
of the Genus by the diffeétion of fome of the larger fpecies.
The part, called by Linneus, the Corol, Mr. Curtis names the
Neétary. Singular appendage of the Seeds of Euphórbia,
taken notice of by Mr. Curtis. 143, Definition of the
term Neétary. Honey profufe in the flowers of the A'r-
butus Unédo; found at the bafe of the petals of Papilionaceous flowers. Clover contains much honey. Chief
diftinétions of thofe Neétaries, which adhere to the parts of
Fruétification. 144, Neétary of Fritilláira, moft obvious in
the fpecies Imperiális, Crown-imperial. Different kinds of
Neétary. Neétary, the term applied by Linneus, to every fingularity of Fruétification, which cannot be reduced under the
feven regular parts of a flower. 145, Neétary, as a feparate
appendage, not found in all flowers. All flowers believed, by
Linneus, to contain honey. Neétaries diftinguifhed, by Linneus, into two kinds. The tube of the Florets of Compound
Flowers contains honey. Neétary only noticed by Linneus
when it charaéterizes a Genus. 146, The tube of One petalled

Flowers

Flowers termed, by Linneus, a true Nectary; he calls the stamens of Fraxinélla, Nectar-bearing. Refinous matter on the
filaments not of the nature of honey; fimilar to that with
which the ftalks abound. Nectaries placed apart from the
Fructification; the ftructure of them merits the ftricteft attention. 147, Nectaries of Colombine refemble the parts of a
bird. Beauty of the Nectaries of Helléborus and Parnáffia;
Globules not the true Nectaries. The bafe of the petals
of Pinks fweetifh. The bafe of the Calyx replete with honey. Difficult to determine by what part of Fructification the
honey is fecreted. Beautiful ftructure of the Nectary of
Mignonette. 148, Structure of Paffiflóra. Nectaries form
the principal feature in the Genus Paffiflóra; in fome fpecies
refemble a bread-bafket. 149, Linnean defcription of Paffiflóra not juft. Difficulty of attaining a diftinct idea of the
Gynandria clafs. Extraordinary ftructure of Fructification
peculiar to the Orchis tribe. 150, Orchis flower diffected.
Twifted Germe of Orchis; curious ftructure of the Stamens,
and the cafes by which they are contained; may be drawn out
of their cafes by the moft gentle touch. Globule at the bafe
of each Stamen. Anthers compofed of Corpufcles; fame effect,
probably, produced by them as by Anther-duft. Seed of
Orchis apparently perfect. 151, Smallnefs of feed no argument againft it's vegetating. Ferns propagated from feed.
Orchifes not yet decidedly fo; increafe fparingly by the root.
Patience and impartiality requifite to make experiments.
152, Early Purple Orchis obvioufly diftinguifhed by it's
fpotted leaves, and brilliant flowers. Orchis Mório appears
under many varieties; marked through all it's varieties by the
green lines on the two outermoft petals; Anthers green. Ten
diftinct fpecies of Britifh Orchis. Different Génera of the
Orchis-like plants diftinguifhed by their Nectaries. Bee-órchis
an Ophrys. 153, Characters of the Ophrys Genus fhould be
examined with magnified drawings. Different ftructure of
Orchis and Ophrys. The character of feveral fpecies taken
from the Nectary. Leaves of Ophrys Apífera, and Ophrys

Ováta,

Ováta, differ materially from the leaves of the Orchis Genus.
Roots of Ophrys Apífera refemble thofe of Orchis. Roots of
Ophrys Ováta fibrous. Suppofed error in the character of the
feed-veffels of Orchis, Satyrium, Ophrys, and Serápias.

LECTURE III.

Génera of the Claffes One-houfe and Two-houfes. Of Ferns.

PAGE 157, Arum, a plant of extraordinary ftructure.
Nature not limited in her modes of reproduction. 158, Singular fituation of the ftamens of Arum, refpecting the Piftil.
Stamens a collection of Anthers only. Nectaries of Arum.
Seeds of Arum. Opinion of the younger Linneus of the claffic
character of Arum. 159, Roots of common Arum extremely
acrid; eaten by thrufhes; the roots of fome fpecies made ufe
of as food; the leaves of fome fpecies boiled and eaten.
Starch made from the roots of Arum Maculatum; injurious
to the hands which ufe it. All parts of the plant acrid. The
leaves and whole ftructure of Hydrócharis exceedingly curious.
160, Singularities of the ftamens explained. Nectaries obferved,
by Mr. Curtis, on the piftil, not noticed by Linneus. Spathes
of the flowers of Hydrócharis appear full of bubbles. Mr. Curtis's account of Hydrócharis differs from that of Linneus.
161, Flowers of Typha, or Cat's-tail, difficult of inveftigation. Mr. Curtis does not wholly agree, in his account of
them, with Linneus. Mr. Curtis's account to be relied on.
Flowers of Typha defcribed. Suppofed calyx, of Linneus,
hairs which cover the receptacle after the ftamens are fallen
off. Spikes of flowers Aments, or Catkins. Cylindric form
of the fpikes marks the Genus Typha *Culm*, the Linnean
term for the ftraw of Graffes. 162, Difference of pofition of
the male and female flowers on the *Culm*. Magnificent appearance

pearance of the flowers of Typha Major; every part of the
plant worthy of attention. Species of Cárex not eafily diftinguifhed from each other. Cárex Péndula diftinctly marked
by the long pendant Aments of it's flowers. The Catkin tribe
of flowers merits attentive examination; manner of inveftigating Ament-bearing plants. 163, Cryptogámia clafs. Stamens and piftils not yet difcovered in the Cryptogámia clafs.
Meaning of the term Fructification, as applied to the plants of
Cryptogámia. The Filices, or Ferns, divided into three fections,
by the difpofition of their fructifications. Radical Fructification explained, well feen in Pilulária. Hedwig's botanical
refearches, in clafs Cryptogámia, of great importance.
164, Equifétum Sylváticum, a good fpecimen of the fpiked
fructification of Ferns. Extraordinary appearance of the fuppofed feeds of Equifétum; magnified drawings a great affiftance in the inveftigation of obfcure plants. Plates not wholly
to be relied on. Little progrefs made in any ftudy by thofe
who rely on the authority of others. The rule " See for your-
felf," to be obferved in all ftudies; Mr. Curtis's works rendered
valuable by the obfervance of this rule. 165, His candid correction of the few errours of Linneus, of effential fervice to
the botanical world. Account of the progrefs of Equifétum.
Greenifh powdery mafs fhook from the fpike. Particles of
powder appear regular formed bodies, viewed in the microfcope; account of their form. Regular organization of the
parts of plants. Curious appearance of the powder fhook
from the fpikes of Equifétum. 166, Hedwig's opinion of
this powder; circumftance in favour of his opinion. Scales of
the protruded fpike of Equifétum, protected the fpikes before
protrufion. 167, Knowledge of the fructification of Equifétum
leads to the knowledge of the Fructification of other fpiked
Ferns. Leafy Fructification: beauty of the maiden hair. The
parts of Fructification too minute for the inveftigation of young
Botanifts. 168, The larger fize of Hart's-tongue, fhows the
Fructification diftinctly. Fructification defcribed; wonderful
mechanifm of the feeds, with their apparatus. Benevolence of

nature

nature in all her works. Mechanifm of the Capfules of Fern.
169, Difficulty of viewing the Capfules of Fern through a
microfcope. Capfules opened by the warmth of the breath.
Have the appearance of being alive; dextrous management,
and patience required in viewing them. Root of Polypodium
Vulgare refembles the large kind of caterpillars. 170, Errour
in the defcription of Polypódium Vulgare by eminent Botanifts;
afcribed by Mr. Curtis to too great deference to authority.
Errour of Tournefort in delineating the Capfules of the Polypódium Genus without rings; one of the many inftances of
the fallacy of authority 171, Polypódium Vulgare appears
deftitute of the membrane by which the Capfules of all the
other fpecies are enclofed. The Fern tribe opens an ample
field of difcovery to modern Botanifts. Practice can alone
make us acquainted with the different Génera of Ferns. Similarity of their Fructifications. Capfules varioufly placed on
the Fronds; precife géneric character not eafily afcertained.
172, Plates and remarks of Mr. Curtis, in his London Flora,
particularly ufeful in the ftudy of Ferns.

LECTURE IV.

Moffes, Flags, and Fungufes.

PAGE 177, Moffes, a tribe of plants little underftood;
beauty and ufe of Moffes. The opinion that they impoverifh
the ground on which they grow, erroneous. Roots of Moffes
penetrate but a little way into the earth. 178, Refemble
Ever-greens. Fuel, called Peat, formed from the roots of Mofs.
Peat-fuel not exclufively derived from Mofs. 179, Whole
trees enter into the compofition of a Peat-bed. Mofs retains
moifture a long time, without becoming putrid; it's ufe to
gardeners. The diftinct Fructifications of Moffes well eftablifhed
                                                              fince

fince the time of Linneus, their fituation not yet determined.
A revifal of the works of Linneus defirable. Clafs Cryptogámia improved fince his time. Génera of Moffes diftinguifhed
by their outer habits, and fituation of their Capfules. Refemblance of Moffes to the Pine tribe; 180, flowness of their
growth. Difference in the leaves of Moffes. Male and female flowers placed feparately. Calyx, termed by Linneus
Calyptre. From the prefence or abfence of the Calyptre
Linneus has diftinguifhed the Génera. Opérculum of Moffes,
a curious microfcopic object; fhould be examined with magnified drawings. 181, Parts of the Fructification of Moffes
may be feen, in an early ftate, with the affiftance of glaffes.
Hedwig's difcovery of the difference betwixt the leaves of the
plant, and thofe which form the fructification buds; efteems
the bud-leaves true involúcres; increafe in fize as the capfules
grow towards maturity. Hedwig's refearches promife great
information on the fubject of Moffes. 182, His plates not of
much ufe to young Botanifts. Mr. Curtis's figures and defcriptions accurate and plain. Mr. Curtis does not venture
to decide whether the powder contained in the Capfules of
Moffes is Anther-duft or Seed. Hedwig afferts that the Capfules are true Seed-veffels. Young Plants raifed from the
Capfules of Moffes, by Hedwig; fowed, by Dillenius, without
fuccefs. Caufe from which thefe different refults of the fame
experiment may have arifen. Defcription of Curled Bryum.
183, Hedwig's obfervation upon the expanfion and contraction
of the Fringe of the Capfule in dry and moift air; clofes, even
from the moifture of the breath. Curious mechanifm of the
Capfule of Moffes; contents of the Capfule protected by the
fringe found under the Calyptre. Calyptre of Bryum Undulátum
defcribed. 184, Mechanifm of the fuppofed Fructifications
of Moffes and Ferns equally curious; both feem formed for
the protection and difperfion of their feeds; the manner in
which the feed is produced unknown, unlefs Hedwig's refearches may be relied on. Magnified leaf of Bryum Undulatum fhows it's undulated edges. Bryum Undulatum

produces

produces it's Capfules from November to February; fituations
in which it is found. The leaves curl up foon after the plant
is gathered. A fpecies of Bryum placed by Linneus among
the Mniums; diftinguifhable from Undulatum by it's bending
peduncles. 185, Star-like appearance on Moffes fuppofed, by
fome authors, to be the piftil-bearing parts of Fructification.
Various opinions refpecting thefe Stars; conjecture refpecting
thefe Stars. An outline of the opinions of eminent Botanifts
on the clafs Cryptogámia fhould be given to botanical pupils; admits only of conjecture. The part, termed Anthers
by Linneus, now known by the name of Capfule. 186,
Singular ftructure of the leaves of Hypnum Proliferum,
found by Linneus under the fhade of thick woods. Rare appearance of Fructification in Hypnum Proliferum. Time of
fructifying, from December to February. 187, Structure of
Capfules nearly the fame in all the Moffes. Peculiarities, difcovered by Mr. Curtis, in the Capfules of Bryum Subulátum
and Polytrichum Subrotúndum. The ufe of thefe peculiarities
not underftood. Great nicety requifite in making experiments.
188, Curious and beautiful ftructure of the Capfules of Polytrichum Subrotúndum difcovered to be a conftant mark of the
Genus. Structure of the Capfules defcribed. 189, London
Flora a work too expenfive for general ufe. Dr. Smith's
Englifh Botany recommended. 190, The Root, Stem, and
Leaf of Algæ fcarcely admit of diftinction. Deftitute of obvious Flowers; manner of diftinguifhing the Génera. Algæ
of great importance in the economy of Nature; vegetate upon
the bareft rocks. Lichen Pafcalis found by Dr. Smith on a
torrent of hardened lava; peculiarly fitted for the beginning
of vegetation on a hard furface. Thread-form Lichens infinuate their roots into crevices of the barks of trees. 191, Cruftaceous kinds vegetate on fmooth furfaces. Procefs of Nature
in forming vegetable mould apparent upon the fmooth and
barren rocks upon the fea-fhore; account of the procefs.
Lichens made ufe of in dying; fed upon by goats and rein-
deer. 192, Cup-mofs, a Lichen. Numerous fpecies of Lichen
difficult

difficult to diftinguifh. Hedwig's inveftigations of them; his opinion of their parts of Fructification. Fringes from Lichen Ciliáris put forth roots; diftinct from the fuppofed parts of Fructification. Hedwig's plates of the Algæ tribe. Algæ not well underftood. Sea-wrack, a Fucus. 193, Prolific property of the leaves of Fúcus Veficulofus. Black hair-like tufts found growing upon Fúcus, a Conférva. Some fpecies of Fúcus, perhaps not true vegetables. Sea-anemóne falfely efteemed a vegetable. Green films on water and on trees not thoroughly underftood. Clafs Cryptogámia requires new arrangement. 194, Géhera of the third order diftinguifhed by no obvious common character; peculiarities of them worth attending to. Beauty of the Lichens. White Mofs; on heaths, Rein-deer Lichen; many varieties of it; diftinction between them and the true fpecies. 195, Mofs on trees, a Lichen. Lichens, Moffes, Ferns, and Fungufes, form a complete winter garden. Fungufes fhould be ftudied with good plates. Generality of Fungufes not offenfive either to the fmell or tafte. Much information gained, concerning them, within the laft twenty years; not yet perfectly underftood. 196, Hedwig's refearches into the Fungi tribe; fuppofed, by him, to poffefs ftamens and piftils. Curtain of Fungufes, not found in every fpecies. Curtain defcribed. Hedwig's account of the fuppofed piftils. 197, Seeds of Fungi. Globules uniformly found in the Génera Agáricus and Bolétus believed, by Hedwig, to be ftamens. Parts which can be feen only with powerful magnifiers cannot be ufed for the diftinction of Génera. 198, Excellence of géneric characters to be obvious and clear. Fungi continue their fpecies by a powder which is vifible in the gills of many of them, generally allowed to be feed. Short continuance of fome of the Agaric fpecies. Inveftigation of an Agaric. Genus Agáricus defcribed. Three firft divifions of the Genus founded on the pofition of the ftipes. 199, Diftinction betwixt the Volvè and Curtain, explained by Mr. Bolton. Erroneous account of the Volve, by Linneus. Under the Curtain of Fungi the parts of Fructification found, by

Hedwig.

Hedwig. Ring of Fungufes formed from the remnants of
the Curtain. Ring uncertain in it's appearance; cannot be
ufed for a permanent mark. Stem of Agáricus either folid or
hollow; varies much in it's degree of folidity. 200, Colour of
the gills varies in different fpecies; vary much in their refpective lengths. Seeds formed between the membranes of the
Gills. Situation of the Gills. Peculiarity of ftructure difcovered, by Mr. Curtis, in the Gills of Agáricus Ovátus; ufe of
that ftructure. Secondary fubdivifions of the Agarics, on
what founded. 201, Gills a part of great importance; various appearance of the Gills; colour of the Gills not liable to
vary. Character of the fpecies taken from the colour and
ftructure of the Gills. Colour changes when the plant begins to decay; colour muft be obferved in their firft ftate of
expanfion; colour of the flat fide of the Gills, that which
muft be attended to. 202, Hat of the Agarics, the part
leaft to be depended on. Vifcous juice of the Hat depends
on the ftate of the atmofphere. Acrid juice in Agarics, not
conftant. Dr. Withering's arrangement of the Fungi. 203, Exception to the uniformity of colour in the Gills in Agáricus
Aurántius. Beautiful colours of the Agarics. Agáricus Cæfareus the moft fplendid of the Agarics; a rare plant in Britain,
common in Italy. Agáricus Campéftris, the Fungus moft
commonly eaten in England; method of propagating it. Caprice of mankind in their choice and rejection of food. 204, All
kinds of Fungi ufed for food by the Ruffians. Doubtful whether the common Mufhroom be poifonous. Many vegetables
rendered wholefome by fire. 205, Neceffitous fituation of the
inhabitants of northern climates. Make ufe of the inner bark
of the Pinus Sylvéftris for food. Method of preparing it for
bread. Swine fattened upon pine-bark bread. 206, Numerous
tribes of infects fuftained by the Fungi. Extenfive ufe of the
Pinus Sylvéftris, Scotch Fir; roots of Scotch Fir ufed in the
Scotch Highlands for candles. Ropes made by fifhermen of
the inner bark. 207, Pinus Sylvéftris the only fpecies of Fir
which grows naturally in Scotland. Oil extracted from the

cones

cones of Scotch Fir; lives to a great age; profuse in Anther-duft. Mould a regular plant; it's parts diftinctly feen through a microfcope. 208, Thirteen different fpecies of the Múcor Genus. Golden Múcor, ftains the fingers yellow, when touched; commonly found on the Genus Bolétus; repels moifture.

LECTURE V.

On the Graffes.

Page 211, The Grafs tribe requires a particular mode of inveftigation. Vague idea conveyed by the vulgar term Grafs. Graffes imperfectly underftood until late years. Names by which they have been diftinguifhed not in general ufe; 212, fubject greatly elucidated by Mr. Curtis; his Practical Obfervations on Britifh Graffes; ufeful knowledge to be ac-quired from that work. Graffes form one of the natural orders of Linneus. Corn arranged under the fame order. Similarity in the parts of Fructification of Graffes. Striking agreement in their outer habits. Whole clafs characterized by fimplicity of ftructure. 213, Seed of Grafs does not divide into lobes when it germinates; termed, by Linneus, One-cotylédoned; the hufk of the feed may be feen adhering to the fibres of the young plants of wheat. 214, Peculiarities of Graffes fhown in Alopecúrus Praténfe, Meadow Fox-tail; better feen in the plant than in plates. London Flora amufing and informing on Graffes. Leaves and fheaths of Graffes often furnifhed with briftles. 215, Specific characters taken from the prefence or abfence of briftles. Parts of Fructifica-tion not noticed by common obfervers. Beauty and ftructure of thofe parts worthy of the higheft admiration. Natural character of the flower of Graffes. Arifta of Graffes. Awn of barley particularly ftrong; not conftant in every fpecies.

Corol

Corol of Graſſes termed *Glume*. Diviſions of the outer Glume often mark the Genus. Difficulty of diſtinguiſhing the Calyx from the Corol. Calyx and Corol to be underſtood according to the definition of Linneus. Nectary of Graſſes diſtinctly ſhown in Mr. Curtis's plates; 216, not difficult to be ſeen in the natural flower. May be ſeen at the baſe of the Germe in Wall Barley; nearly reſembles the Corol; furniſhes no generic diſtinction. Three ſtamens, the number commonly found in Graſſes. Two piſtils. Exceptions to this number. Styles beautiful; ſeen with advantage through a microſcope. 217, Cloſe-ſpiked Graſſes do not ſhow their Fructification well. Seen well in Feather-graſs. Should be examined before the Anthers have diſcharged their duſt. The flowers of Graſſes have no ſeed-veſſels. Seeds emitted from the Calyx in various ways. Seeds of Feather-graſs diſperſed by the twiſting of their Awns. Receptacle of Graſſes. The Stem lengthened out. Awns of Feather-graſs twiſt after they have been gathered. 218, The parts of Fructification obvious in Quakegraſs, Bríza Máxima. Characters of Fructification nearly conſtant in Graſſes of the Triándria claſs. Strict adherence of Linneus to the claſſic character of Graſſes. Hólcus Lanátus placed in the claſs Polygamia. 219, Greatneſs of the works of Linneus a juſt excuſe for the few errours contained in them. Variation of the number of ſtamens not uncommon in ſeveral ſpecies of Graſs. Strict adherence to the claſſic character perhaps advantageous in an arbitrary ſyſtem. Anthoxánthum judiciouſly placed in the claſs Diándria from it's conſtant number of two ſtamens. 220, Peculiarities in the Fructification. 221, Fragrant ſcent of hay derived from the leaves of Anthoxánthum; not the only Engliſh Graſs which is fragrant. 222, Flowers of annual Póa ſaid to be ſo by Mr. Swayne. Anthoxánthum, viviparous; many Alpine Graſſes viviparous. Canary birds fed on the ſeeds of Phálaris Canariénſis. Ribbongraſs, a ſpecies of Phálaris. Genus Avéna, marked by the twiſted awn on the back of the corol. 223, Motion of Avéna Fatua. Named Animated Oat. Curious circumſtance reſpect-

ing

ing the feed of barley. Automaton ingenioufly made on the principles of the awn of barley. Wheat the moft nutritive of the grains ufed for food; found in moft parts of Europe and of Afia. 224, Zéa, Indian Wheat, the product of the torrid zone. Rice of the natural order of Graffes; feparated from them in the artificial fyftem of Linneus; chief food of the inhabitants of moft eaftern climates; converted into poifon by the fpirit extracted from it. Extenfive utility of the natural order of Graffes; their roots not deftroyed by being trampled upon. The Flowers of plants not eaten by cattle. 225, Admirable provifion made by Nature for the prefervation of Graffes.

LECTURE VI.

Specific Diftinctions and Double Flowers.

Page 227, Linneus firft began to form effential fpecific diftinctions of plants. Confufion arifing from the want of fuch diftinctions. Specific diftinctions of Linneus. 228, Trivial name, given by him, generally arbitrary; refembles the name given to the individuals of a family; advantage of fuch names in preference to defcriptive names. Confufion arifing from the neglect of the ufe of proper names. Perfection of Nomenclature may be hoped for. 229, Great advantage of the ufe of the proper names and the terms of fcience. Excellence of the language of the Lichfield tranflation of the Syftem of Vegetables. Awkwardnefs of forming Englifh trivial names. Such names injurious to the fcience of Botany; 230, defended only by fuperficial Botanifts. Specific characters not to be formed from variable circumftances. Colour one of the leaft permanent characters. 231, Departure of Linneus from his own rule. Roots of plants a true fpecific mark.

Difficulty

Difficulty of examining the root prevents it being made ufe of
as fuch. Trunk and Stalk afford ftrongly marked characters.
Fulcra and Inflorefcence furnifh permanent marks. Parts of
Fructification fometimes ufed with advantage in fpecific diftinctions. 232, Some Hypéricums and Gentians diftinguifhed
by their parts of Fructification. Such diftinctions agreeable
from being obvious. Many other fpecific characters equally
obvious. Study of leaves neceffary to the underftanding the
fpecies of plants. Moft elegant fpecific diftinctions formed
from leaves. Great variety in leaves; muft be attentively
ftudied; method of ftudying leaves. 233, Form of leaves firft
to be confidered; divided into fimple and compound; fimple
leaf defined; fixty-two ways in which a fimple leaf may be
diverfified. Various forms of leaves muft be ftudied with plates
of them, and terms of explanation. Genius of Linneus fhown
in the conftruction of his botanical language. Englifh Botanifts much indebted to the Lichfield tranflators of Linneus's
works. Preface and advertifement to the Lichfield tranflation
fhould be read by botanical pupils. The knowledge of leaves
may be acquired by attention. 234, Explanation of the
Linnean language. Excellence of the Linnean defcriptions.
Want of precifion in the defcriptions of other authors. Method of acquiring precife ideas of the different forms of leaves.
235, Language of the Lichfield tranflators explained; agreeable concifenefs of that language. 236, Compound leaf defined. Compound leaf and branch known from each other by
two rules. 237, Leaves of Robinia Pfeud-acacia, a good example of the compound character. Three kind of compound
leaves. Great variety of compound leaves. Each modification
of a compound leaf marked by an appropriate term; method
of ftudying compound leaves. Different modifications of the
compound leaf enumerated. Fingered leaf feen in Horfe-
chefnut and Lupine. Specific characters frequently formed
from the various modes of compound leaves. 238, Various
forms of fimple leaves fhould be ftudied before thofe of the
compound kind are attended to. The Lichfield tranflation

the

the only book from which an Englifh Botanift can completely learn the fcience of Botany, Determination of leaves explained. Belongs to fimple and compound leaves equally. *Alternate* leaves fhown in Ivy-toad flax. 239, *Oppofite* leaves, in Myrtle. Manner of leaves being placed on the ftem common to the whole Genus. *Direction* of leaves explained. Various modes of direction muft be ftudied. *Infertion*, a general term for the manner in which leaves are attached to plants. Each mode has an appropriate term; thefe terms well explained in the Syftem of Vegetables. Double flowers, fome knowledge of them requifite for young Botanifts. 240, Double flowers, the pride of florifts, the product of culture. Vulgar errour of gardeners refpecting double flowers. Completely double flowers lofe their ftamens. Various modes of vegetable monfters being produced. Calyx and lower row of petals unchangeable in double flowers. Half-double flowers bear fruit. 241, Hofe-in-hofe Polyánthos, a proliferous flower. Hen-and-chicken Daifie, a beautiful vegetable monfter. Extraordinary change caufed in Rofe Plantain, by becoming double. Flowers multiply by their nectaries; become double in various ways. Provence Rofe deftitute of ftamens. Damáfk Rofe does not lofe it's ftamens by becoming double. Many-petalled flowers moft liable to become double. One-petalled flowers rarely multiplied beyond a double corol. Beauty of compound flowers increafed by multiplying. Single flowers generally more beautiful than double ones. 242, Various caufes from which plants depart from their true fpecies; culture the moft prevailing caufe. Fruits and efculent vegetables derive their excellence from the art of gardening. Culture the beft teft of a true fpecies. Ingenuity and induftry of mankind confpicuous in the culture of corn. Botanifts fhould attend to diftinctions arifing from feedling varieties. Varieties of plants not noticed in the Syftem of Vegetables, marked in the Species Plantárum with a capital B. Leaves fubject to all the varieties which take place in flowers; 243, undergo extraordinary changes in their appearance. Many changes in leaves may be effected by art.

NOTE.

NOTE.

IN the pronunciation of the names of plants, *e,* at the end
of Latin and Greek words is always pronounced, and not funk
as in Englifh. Thus, Agáve, is pronounced A-gá-ve; and
Acre, A-kre.

Ch in thefe languages is pronounced like *k* in the Englifh.
Thus, Achilléa is pronounced as if it were fpelt A-kil-le-a;
and Chelóne, as if it were fpelt Ke-lo-ne. In words ending
in *ides,* the *i* is always to be pronounced long. In words
beginning with *fce* and *fci,* the *c* is generally pronounced foft.
In words from the Greek, the *g* fhould be pronounced hard, as
in Syngenéfia and Storgé.

BOTANICAL LECTURES.

PART THE FIRST.

LECTURE I.

The Seven Parts of Fructification explained.

LINNEUS, the great fwedifh naturalift, has divided the vegetable world into 24 *claffes*; thefe claffes into about 120 ORDERS; thefe orders contain about 2000 *families*; and thefe families about 20,000 *fpecies*, befide the innumerable varieties, which the accidents of climate or cultivation have added to thefe fpecies. The fyftem of Linneus is called the fexual fyftem of botany, from being founded on obfervations, which feem to prove, that there are males and females in the vegetable world, as well as in the animal. The ftamens are termed males, and the piftils females: thefe moft frequently exift in the fame flower,

B   but

but are fometimes in different flowers, and
fometimes even on different plants; and from
their number, fituation, and other circumftances belonging to them, he has formed his
claffes and ORDERS; his *families*, or *genera*,
are formed from all the parts of the bloffom
or fructification; his SPECIES, which are individuals of the families, from the leaves of
the plant; the varieties, from any accidental
circumftance of colour, tafte, or odour: the
feeds of thefe varieties do not always produce
plants fimilar to the parent, but frequently fuch
as refemble that fpecies to which the parent
belonged. Having given a fketch of the phi-
lofophy of the fyftem, the next thing to proceed to is the examination of the different
parts of a bloffom, or, according to Linneus,
the fructification. Nor is a knowledge of any
other than the englifh tongue neceffary to
the acquirement of the language of botany:
the latin pupil may know that the word calyx
fignifies cup, but that will not affift him in
the knowledge of the various fpecies of calyxes which he will have to retain in his memory; the common meaning of words is not
fufficiently precife for the purpofe of fcience,
and cup and calyx require equal explanation
when

(3)

when appropriated to the particular parts of a flower. The works of Linneus being now tranflated, botany has a language peculiar to itfelf; that language is, perhaps, fomewhat lefs difficult to learn than any other language; and fhould tenfold the difficulty be found in the acquirement of it, the time might be esteemed well fpent.

The term fructification is defined by Linneus to be a temporary part of vegetables dedicated to germination; that is, all the parts of the bloffom, which are intended for the production and prefervation of the feed, and which, having brought that to perfection, wither and fall off. All thefe parts, however, are not effential to the production of perfect feed, as will be feen hereafter; nor are all thefe parts prefent in every flower. There are *feven parts* of *fructification*. 1ft, the *calyx*; 2d, the *corol*; 3d, the *ftamen*; 4th, the *piftil*; 5th, the *pericarp*; 6th, the *feed*; 7th, the *receptacle*. The *calyx* is the termination of the outward bark of a plant; of it there are feven kinds; it generally appears in the form of a green cup; it's chief ufe is to enclofe, fupport, and protect the other parts of the fructification. The firft and moft common kind of calyx is the Perianth,

B 2   and

and is placed immediately under the flower,
which is enclosed in it, as in a cup; primrofes
(prímula) and rofes (rofa) have their calyxes of
the Perianth kind. 2d, Invólucre, which is a
calyx, growing at a diftance from the flower.
Moft flowers which have Invólucres have alfo
Perianths, as the prímula genus. Thofe flender leaves, which grow at the bafe of the
numerous flower-ftems of the polyanthos
(which is a prímula) are termed Invólucres;
the fame in meadia dodecátheon, in parfley,
apium, and all that tribe of plants which is
termed umbelled. The plant called fool's
parfley, æthúfa, by eating of which, miftaking
it for garden parfley, fome perfons have been
faid to be poifoned, may be diftinguifhed from
all other umbelled plants by the Involucres,
which belong to the fmall umbels, and which
confift of three long, narrow, pendulous leaves,
placed at the bottom of each umbel: thefe
leaves are called partial Involucres; thofe which
grow at the bafe of the whole collection of
umbels form what is termed the general Involucre. 3d, Glume chiefly belongs to graffes,
and confifts of one, two, three, or more valves,
folding over each other like fcales, and frequently terminated by a long ftiff-pointed
prickle,

prickle, called the Awn, or beard. 4th, Ament
is, what is commonly called a catkin; it confifts
of a great number of chaffy fcales, difperfed
along a flender thread, or receptacle, and has
obtained the name of catkin from it's fancied
refemblance to a cat's tail. Thefe Aments
are compofed of both male and female flowers;
the Aments or Catkins of the willow-tree, falix,
diffufe a fragrant odour around them in early
fpring; the yellow ones, well known to children by the name of Goflins, from their fancied refemblance to that little animal, contain
ftamens only, and derive their bright yellow
colour from the prolific duft of their tips or
Anthers. The green catkins are the female
Aments, and, when mature, have the appearance of fmall tufts of wool, which is caufed
by the downy material with which their feeds
are crowned. The female *Aments* of Birch,
Bétula, are beautiful, being compofed of ftamens with bright crimfon *Anthers* furrounded
by pale green fcales; the female bloom of
Nut-trees is alfo of an elegant conftruction,
though fo minute as to efcape general obfervation. The 5th fpecies of calyx, called a
Spathe, wraps round the flower or flowers
contained in it, till they are ftrong enough no

longer

longer to require it's protection, and then they burst forth. Sometimes the Spathe consists of one piece, as may be seen in the snow-drop, galánthus nivalis, and daffodil, narcissus, pseudo-narcissus, and in most plants which have this kind of calyx; sometimes of two, as in the Japan lily, amaryllis formosíssima; and sometimes of many. Calyptre is the term for the calyx of mosses. Calyptre is defined by Linneus to be the cowled calyx of moss, covering the anther; which definition strongly expresses this species of calyx; it may, however, be necessary to give some more familiar idea: the calyptre resembles a very small extinguisher of a candle, which covers the flower of moss, and protects it's dust, or seed, from injury: in Mr. Curtis's London Flora there are a variety of beautiful specimens of this kind of calyx; and, in the months of November and December, it may be found growing on every bank. The 7th and last species of calyx is the *Volve*, the term used by Linneus for the calyx of Funguses, a tribe of plants which requires much elucidation, and, joined to some other families of equally obscure habits, form a class confessedly little understood.

The second part of fructification is the Corol,

rol, or that part of the flower which moſt attracts our notice, conſiſting generally of beautifully coloured leaves. Linneus defines it to
be formed from the inner rind of the plant,
as the Calyx is from the outer; it's leaves are
called Petals, a term which ſhould be remembered, as it is neceſſary to prevent confuſion
betwixt the green leaves of the plant, and the
coloured ones of the flower. By the number,
diviſion, and ſhape of the Petals, the different
kinds of Corols are diſtinguiſhed; a Corol is
called one-petalled, when it conſiſts only of
one piece; two, three, or more petalled, according to the number of pieces of which it is
compoſed. The flower of common Polyanthos is one-petalled, although, on the firſt
view, from its diviſions round the margin, it
appears to conſiſt of five petals. The beſt
way of knowing, whether a flower conſiſts of
one or more petals, is to attempt to take them
off all together. The one-petalled flowers, be
their diviſions ever ſo deep, have their petals
united together at the baſe, forming a tube,
ſometimes very long, as in Polyanthos, or
very ſhort, as in Verónica. In flowers of
many petals they are fixed by the claw to different parts of the fructification, which circum-

B 4   ſtance

ftance is frequently of ufe in diftinguifhing one flower from another. Linneus has availed himfelf of it in his formation of the génera, or families of plants. The various fhapes of the corol are alfo of great ufe in this particular, and therefore fhould be accurately underftood*. There are feven different forms of the corol: bell-form, of which there are great varieties; funnel-form; falver-form; wheel-form; crofs-form; gaping and grinning corols, which may be confidered as different kinds of the fame form; and papilionaceous, or butter-fly-form, which belongs to the pea-bloom, or lupine tribe of flowers. There is an eighth form, which does not belong to any of thefe that I have mentioned, and is properly called an irregular flower; of this kind are the monkf-hood (aconítum napellus), violet (víola), lark-fpur (delphínium), orchis, and fraxinella (dictámnus). Campánula is an inftance of the bell-form; of the funnel-form, henbane (hyofcy'amus,) and oleander (nérium); of the falver form, periwinkle (vínca); of the wheel-form, mullein (verbáfcum), and pimpernel (anagállis); the crofs-form may be feen in

* See Plate the Second.

wall-

wall-flower (cheiránthus), and in candy-tuft
(ibéris), and confifts of four petals nearly
equal, and fpread at the top upon claws, the
length of the calyx, in form of a crofs. The
butterfly-form is feen in pease; the gaping and
grinning, in white archangel (lámium), and
fnap-dragon (antirrhínum) There is another
part of the fructification, which Linneus con-
fiders as belonging to the corol, and to which
he firft gave a name; this is the Nectary; fo
he has called that part wherein the honey is
found, from the fancied refemblance to the
fabled liquor of the gods. The nectary frequently forms a part of the corol, but as frequently is diftinct from it: the delicious
juice, whence it derives it's name, is found in
abundance at the bafe of the tubes of the
flowers of honeyfuckle (lonicera), and cowflip (prímula), and equals the pureft fugar in
the richnefs and fweetnefs of it's tafte. A moft
effential part of fructification is the *ftamen*;
as by it the fine duft, or powder, is prepared,
by which the feeds are to be fertilized, and
rendered capable of producing young plants.
The Stamen confifts of three parts; the Filament, the Anther, and the Duft. The Filament is the thread on which the Anther
grows;

grows; the Anther is that part which is open, ignorantly called the feed; it contains the Duft, and, when ripe, opens and fcatters it abroad, for the ufe to which naturē has def-tined it. Clouds of this duft may be feen about Nettles, Urtíca, at their time of flower-ing, and Sweet Gale, Myríca. Nature has guarded, with nice care, this precious duft, as on it's prefervation depends the continuation of the fpecies. The apparatus, by which in many flowers it is defended from injury, is very curious, and often gives a fingular appearance to the corol. In wet years it fometimes hap-pens, that the excefs of moifture caufes the anthers to burft, before their contents are ripe, and thus we lofe our cherries and apples. It has been fuppofed, that the anthers were pre-ferved from injury in rainy feafons by a fine waxy fubftance enclofing their contents. This idea was fufpected, by Reaumur, to be erro-neous fome years ago, and the experiments of the late Mr. John Hunter confirm his opinion. Mr. Hunter affirms, that the fubftance gathered by bees from the anthers of flowers is not wax, as is generally fuppofed, but that it is collected by them as food for the bee-maggots, and forms what is called the Bee-bread. A

part

part no lefs important than the Stamen is the
Piftil, as it contains the feed which receives
it's fertilization from this duft. The Piftil
alfo confifts of three parts, the Germe, the
Style, and the Stigma. Germe is the term
for that part which contains the feeds in their
embryon ftate; when mature, the fame part
takes the name of Pericarp. The Style is that
fmall pillar which grows from the Germe, the
top of which is the Stigma. The Stigma is
a part of great importance, as it receives the
Duft of the Anthers, and conveys it's effence
through the fine veffels of the Style to the
feed contained in the Germe. Indeed the Anther and Stigma are by Linneus efteemed the
effential parts of a flower, and in the ftrict
language of botany they conftitute one; thefe
parts being prefent are fufficient to the pro-
dúction of fruit; without them there can be
none: the prefence of the Stigma implies that
of the Germe, as the prefence of the Anther
does that of the Duft. There is, however,
another part, which the inveftigations of a late
celebrated philofopher feem to make of equal
importance; this is the Nectary. From his
obfervations it appears, that the honey contained in this part is intended by nature for
the

the nourifhment of the Anthers and Stigmas;
confequently, whenever thefe are found, it
will be found alfo; and, although fome flowers
have been faid to be deftitute of it, this affer-
tion may have arifen from want of fufficient
inveftigation, as the part in queftion was fo
little known before the time of Linneus, that
it had not even obtained a name; and we
have yet to acquire the certain knowledge of
it's ufe.

There are eight different kinds of Pericarp, or Seed-veffel: 1ft. Capfule, 2d. Silique,
3d. Legume, 4th. Follicle, 5th. Drupe,
6th. Pome, 7th. Berry, 8th. Strobile. Capfule is a little cheft or cafket, a dry hollow
feed-veffel, when ripe, which fplits in different
ways, and difcharges it's contents, fometimes
with great force, fo as to difperfe them to a
confiderable diftance; an inftance of which
may be feen in the feed-veffels of the different fpecies of Balfam; and, from the violent
manner in which their feeds are ejected from
the capfules, when mature, Linneus has named
the genus, or family, Impátiens. The feed-
veffel of víola, violet, and panfie, is a Capfule;
before this fpecies of feed-veffel is ripe, it is
frequently flefhy and fucculent, like a berry,
which

which pulpy fubftance may probably be intended for the nourifhment of the young feeds. Silique is a Pericarp of two valves, which varies in fize and figure, fome being long and larger, others round or broad and lefs. From their different forms Linneus has diftinguifhed them into Silicle and Silique, and on this diftinction has founded the Orders of one of his claffes: of the Silicle, which is roundifh, the feed-veffels of Allyfon of Crete, Aly'ffum Saxátile, furnifh an inftance, and alfo thofe of Candy-tuft (ibéris); the common wall-flower (cheiránthus), and cabbage (bráffica), are examples of the Silique. The Legume is diftinguifhed from the Silicle and Silique by the manner in which the feeds are fixed to it's edges; in the Silicle and Silique the Seeds are placed alternately on each fide of their futures, in the Legume they are fixed on one fide only; the Silique feed-veffels belong to the cròfs-form flowers, the Legume to the papilionaceous; and it is this part that we eat of french-beans, and of fome kind of pease. *Follicle* is a bag that opens on one fide, which circumftance forms the diftinction betwixt the *Follicle* and the *Legume* and Silique feed-veffels; Periwinkle, *Vinca,* and Swallow-wort,
Afclépias,

Afclépias, have their feed-veffels of the Follicle
kind, which, when the feeds are ripe, open
lengthways on one fide. Drupe is a Pericarp, or feed-veffel, that is generally fucculent or pulpy, having no valve or external
opening, and generally contains within it's
fubftance a ftone or nut, within which lies a
feed, commonly called a kernel: there are,
however, exceptions to this definition; all the
ftone-fruits are properly Drupes. Pome belongs
to thofe fruits which contain within their flefhy
pulp the other kind of feed-veffel called Capfule; the apple (pyrus) is an inftance of the
Pome: the core of the apple is the Capfule;
the pippins contained within the Core are the
feeds; this kind of Pericarp, or feed-veffel, has
no valve or outward opening. What is erroneoufly called the bloffom of the apple was the
calyx. Berry is a pulpy fubftance containing
feeds, difpofed promifcuoufly through the pulp,
without other covering; rafberries (rúbus),
ftrawberries (fragária), goofeberries (ríbes),
anfwer well to this definition: in many génera, or families, the berry and the drupe
feem to have been imperfectly defined. And
here it is neceffary to obferve, that there are
fome defects in this moft ingenious fyftem of
Linneus,

Linneus, which may perplex the pupil in botany; who, however, when early apprifed of them, will not find his progrefs much retarded by the difficulties which they may place in his way: a full ftatement of thefe defects will be found in Mr. Milne's Botanical Dictionary, a book which fhould be in the hands of all young botanifts, as much information may be derived from it; but it is to be lamented, that the author, inftead of pointing out the errors of the Linnean fyftem with the candour due to a work of fuch great ability, has marked the fmalleft failings with a moft ungenerous acrimony. The Strobile is defined to be formed of an *Ament* with hardened fcales, and in common language is known by the name of Cone, or Fir Apple. The Strobiles of the Larch, Pinus Larix, are peculiarly beautiful in their early ftate of growth in fpring, their colours being a mixture of tender green and bright crimfon. The Strobile is the kind of feed veffel found in all the Fir tribe.

The feed is defined, by Linneus, to be the rudiment of a new plant: a Seed confifts of, 1ft, the part which is to become the new plant, and, 2d, of nourifhment for that new plant till it has attained fufficient

ftrength

ftrength to provide for itfelf: the young plant confifts of what are termed the *Plume* and the Radicle; the Plume rifes into the air, and conftitutes the trunk and branches; the Radicle penetrates into the earth, and forms the roots. The Cotyledons, which are the mealy fubftance of the feeds, are converted into a fweet juice by the growth of the plant, and are gradually abforbed by it; from thefe fweet ftores of nutriment, the infant plant draws fuftenance, until, by having put forth roots, it has acquired the power of collecting food from the earth; as lambs, and the young of the higher order of animals, fuck the milk of their maternal parents until they have attained fufficient ftrength to feek abroad for their nourifhment. The Plume, the Radicle, and the Cotylédons, may be well feen in a garden-bean, vicia faba, and fhould be accurately compared and examined with the fame parts in the feed of cucumber, of which a drawing is given in Plate the Third. By laying an almond kernel in water till it is well foaked, and afterwards fplitting it, there may be feen within the lobes, or cotylédons, two fmall leaves, diftinctly formed, beautifully ferrated round their edges, and elevated upón

a little

a little foot-ftalk, which is the *Radicle* of the feed, as the leaves are the *Plume*. If the Cotylédons of a bean be cut off, the young plant, being deprived of nutriment, is ftarved and dies, or becomes very weak; grafs has it's Cotylédons under the ground, which preferves them from deftruction; fo has corn, which, however, is not fafe from all enemies; the wood-pigeon digs with her bill till fhe finds the Cotylédon of the corn, and then eats it, pleafed, probably, with the fweet tafte it has acquired in the procefs of germination as the Plume has fprouted. The care taken by nature for the prefervation and difperfion of feeds is admirable: in fome plants fhe has wrapped them in foft down; as, for inftance, in Cotton Plant, Goffypium; the part from which our muflin dreffes are made having originally formed the foft cradle of feeds; as the material, of which our filks are made, was the cradle of an infect. Some feeds are nourifhed and kept warm by the pulp of our fruits; others are protected by foft hairs: in thiftles (cárduus) they lie in a foft filklike fubftance, the down of the feed of artichoke (cy'nara) is particularly beautiful; others are furrounded by what is termed an Aril. In

C   the

the definition of this term Linneus has de-
parted from his ufual accuracy; he has defined
the *Aril* to be, " the proper exterior coat of
" the feed," from which it is evidently
wholly diftinct, and rather may be faid to
form a part of the Pericarp, or feed-veffel,
than of the feed itfelf. In Fraxinella, Dictámnus, the Aril is very confpicuous, being
compofed of a material refembling parchment,
and is found lying within the fweet-fcented
outer-hufk of the Capfules. In wood-forrel,
Oxális acetosélla, the Aril is a little white cafe,
which, if held in the hand till warm, burfts
with confiderable force, and the fmall fhining
black feeds leap from their coverings with
furprifing velocity. Nature has not been more
various in her modes of protecting the different kinds of feeds from injury during their
infant ftate, than fhe has been ingenious in the
means fhe has contrived for their difperfion,
when arrived at an age of maturity. Some fhe
has enabled to fly by a fmall light crown fixed
on their tops, others have fingle feathers, others
fmall feathery tufts: every child is well acquainted with the feathered feeds of dandelion
(leóntodon), and has proved, by blowing on
them, how fmall a degree of air is required
for

for their difperfion, when ripe. Some have an appendage like a wing, as the feeds of fycamore (ácer); one of the fpecies of centaurea has a feed furnifhed with a tuft fo nearly refembling a camel-hair pencil, that it might be miftaken for one; feather-grafs (ftípa) has a beautiful plume; one of thefe plants makes an elegant appearance, when in a bright day, with a gentle wind, a number of thefe plumes are feen together, waving in the air, and fhining like filver. But the moft curious of the flying feeds is that of the tillándfia: this plant grows on trees, like the mifletoe (vifcum), and never on the ground; the feeds are furnifhed with many long threads on their crowns, which, as they are driven forwards by the winds, wrap round the arms of trees, and thus hold them till they vegetate: this is very fimilar to the migration of fpiders on the goffamer, who are faid to attach themfelves to the end of a long thread, and rife thus to the tops of trees or buildings, as the accidental breezes carry them. Thefe flying feeds are carried to a very confiderable diftance from their parent plant; others have hooks, by which they attach themfelves to the hair or feathers of animals, or a glu-

tinous

tinous fubftance, in which the feed is lodged, as mifletoe. The feeds of aquatic plants, and thofe which grow on the banks of rivers, are carried many miles by the currents, into which they fall; fome of the American fruits, among which is the cocoa-nut (cócos), are annually thrown on the coafts of Norway. Some account of thefe emigrant feeds, with fome beautiful lines to which this wonderful fact has given rife, may be feen in the Botanic Garden*, a book which contains fuch variety of knowledge, on the fubject of botany, and that knowledge fo diftinctly and agreeably given, that there cannot be one from which more information or amufement can be de-rived.—Birds are the means of difTeminating fome kind of feeds, either by dropping them as they carry them from place to place, or by parting with them whole, after they have fwallowed them. In this way feeds are fre-quently dropped in the hollows of trees, in which fituation, if they meet with a fufficient quantity of foil and moifture, they vegetate, and make an extraordinary appearance, form-ing an union of two diftinct fpecies. A

* See Part the Second, p.128, l. 411.

Mountain-

Mountain-Ash, thus engrafted betwixt the branches of an Apple-tree, is now growing in my garden, and continues yearly to increase in size and vigour, exhibiting a striking contrast to the old decaying tree by which it is supported. It is not exactly known in what manner such trees receive their nourishment; probably they become parasite plants, and derive their food from the juices of the tree to which they are attached, or, perhaps, live chiefly on the air, as those trees must necessarily do, which grow in the fissures of rocks or walls, where there is not earth sufficient for their sustenance. Lastly, seeds are persed by an elastic force in the seed-vessel, or in some part belonging to the seed Stipa (feather grass), as it's seeds arrive at maturity, dislodges them, by twisting the base of the long feather by which they are crowned, till it detaches the seed from it's receptacle, and carries it to a considerable distance from the plant: thus are the seeds of Geranium and Oat dispersed by the twisting of the Awns which crown them.

The Receptacle is the last part of fructification that is to be considered, by which all the other parts of fructification are con-

nected,

nected, and by which they are supported: it is called a proper receptacle when it supports the parts of only one flower, as in prímula, anemóne, and tulip; a common receptacle, when it supports several florets. This last kind of receptacle belongs to what are called the compound flowers, an explanation of which must be deferred until those plants come under confideration. An inftance of a *common* receptacle may be feen in fcabious (fcabióſa), dandelion (leóntodon), and daify (bellis); all thofe parts, which appear to be the leaves of one flower, are perfect flowers themfelves. And here I recommend to my pupils, whether children or adults, to acquaint themfelves intimately with the feven parts of fructification, and with the various fpecies of Calyx, Corol, Pericarp, and Seed, as defcribed in this firft lecture; which may be effected by comparing the different parts of natural flowers with the drawings given of them in Plates Ift, IId, and IIId.

EXPLANATION

EXPLANATION OF PLATE I. PART I.

OF THE SEVEN PARTS OF FRUCTIFICATION.

Fig. 1. The parts of FruCtification of a Crown imperial. Fritillaria imperialis.

    a, a, a, a, a, a, The Petals.

    b, b, b, b, b, b. The Stamens.

    c, c, c, c, c, c. The Anthers.

    d. The Germe.

    e. The Style.

    f. The Stigma.

Fig. 2. A Petal and Stamen of Crown imperial. *g,* the Nectary. *h,* the Anther ſcattering it's Duſt.

Fig. 3. The Pericarp of Crown-imperial cut acroſs to ſhow the three Cells.

Fig. 4. The Perianth of a Roſe, *i, i, i, i, i.*

Fig. 5. The Involucre of Prímula, *k, k,* with the Perianth of the ſingle Flower, *l.*

Fig. 6. A Flower of Graſs. *m,* the Glume. *n,* the Stamens, *o,* the feathered Stigmas of the Piſtils.

Fig. 7. A Male Ament, containing the Stamens only.

Fig. 8. A Female Ament, containing the Piſtils only.

Plate 1. *Part I. P. 24.*

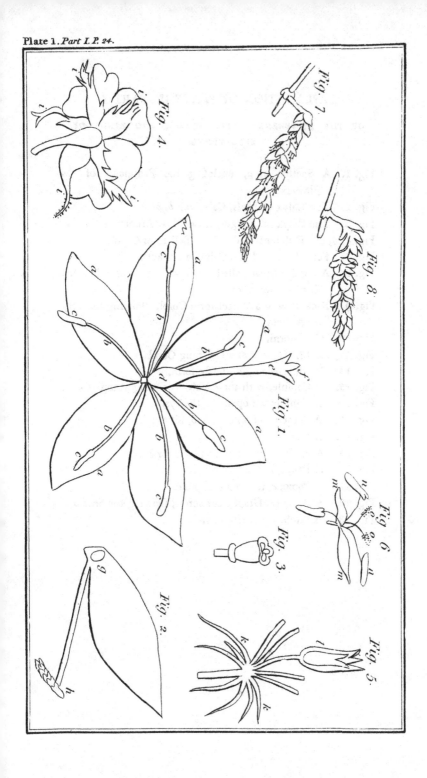

London, Published May 1, 1797, by J Johnson, S'Pauls Church Yard.

EXPLANATION OF PLATE II. PART I.

OF THE DIFFERENT SHAPED COROLS AND KINDS OF SEED VESSELS.

Fig. 1. A Spathe, *a, a,* enclosing the Peduncles of the Flowers.

Fig. 2. The Calyx of Mofs, Calyptre, *b, b.*

Fig. 3. The Calyx of Fungus, *c,* called by Linneus a Volve.

Fig. 4, 5, 6. Different kinds of the Bell-form Corol.

Fig. 7. Funnel-form, *d,* the Calyx, a Perianth.

Fig. 8. A regular one-petalled Corol with a long tube, the Corol Salver-form.

Fig. 9. Back view of a Wheel-form Corol, fhowing the very fhort tube.

Fig. 10. Crofs-form.

Fig. 11, 12, 13. Gaping and Grinning Corols.

Fig. 14. Papilionaceous, Butterfly-form.

Fig. 15. A Capfule, with three Valves opening at top, *a, a, a;*

Fig. 16. A Capfule cut open lengthways.

Fig. 17. A Silique and Silicles, *b. b,* Silicles.

Fig. 18. A Legume.

Fig. 19. A Follicle, with it's receptacle for Seeds, *c.*

Fig. 20. A Drupe, *d,* the Stony Seed.

Fig. 21. A Pome, *e,* the infide Capfule.

Fig. 22. A Berry (A Grape) cut acrofs, fhowing the Seeds.

Fig. 23. A Strobile, cut lengthways.

Plate 2. Part I. P. 26.

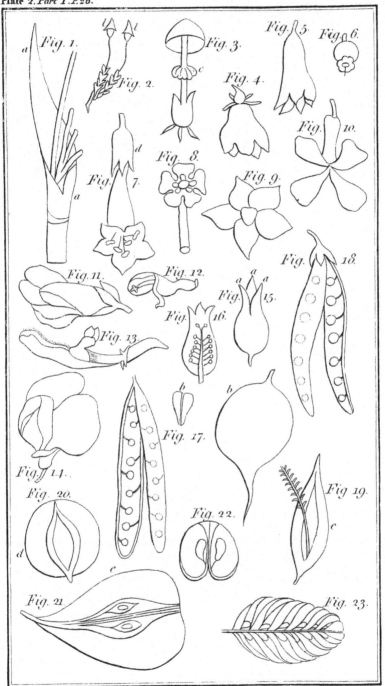

London. Published May 1st 1797. by J. Johnson St Pauls Church Yard.

LECTURE II.

A Flower diſſected: the different kinds of Fulcra and Infloreſcence explained.

THE ſeven parts of fructification, with all their varieties, being well underſtood, the diſ-ſection of a few flowers will be both amuſing and inſtructive. The Verónica and Crowfoot are plants which may be found near every houſe, and afford ſpecimens of the Perianth kind of calyx; the earth-nut (búnium) is an inſtance of the Involucre, and at the ſame time the ſingle florets ſhow the *Perianth*, although ſo very minute that it is liable to eſcape the notice of common obſervers. The male bloom of walnut (júglans) ſhows the Ament; the narciſſus the Spathe. The other three kinds of calyx, the Glume, the Calyptre, and the Volve, as they belong to peculiar and difficult claſſes of plants, would at preſent only per-plex; the ſtudy of them will be therefore better deferred till the pupil is farther ad-vanced in his knowledge of botany.

7   The

The verónica and hare-bell, hyacinthus non fcriptus, have the appearance of many-petalled flowers; but if the corols are taken with care from their receptacles, they are found to confift of one piece flightly united at the bafe. In the hare-bell and verónica we have inftances of the bell-form and wheel-form corols, although the wheel-form of the verónica is lefs decided from the inequality of the breadth of the divifion of it's petals, the lower divifion being narrower than the three upper ones; which nice circumftance is made ufe of by Linneus to diftinguifh this family from all others to which it bears any refemblance. The curling divifions of the corol of the hare-bell difguife it's form alfo; but in neither of thefe génera is the form of the corol the effential character of the family; and is therefore of lefs importance. The Genus of crowfoot (Ranúnculus) is difcriminated by an appearance equally minute as that of the verónica; a fmall protuberance at the bafe of the inner part of each petal being found in every individual of the ranúnculus tribe, even in the double flowers, affords a marked cha-racteriftic of that family. The minute cir-cumftances, of which Linneus has availed
himfelf

himſelf in the diſcrimination of one plant
from another, fills us with admiration; till his
time there was much confuſion in the ranunculus tribe ; his penetrating eye marked this
ſmall appendage to the petal, to which he has
given the name of Nectary ; he found it to
exiſt uniformly in the individuals of the genus ; and we are now no longer at a loſs to
diſtinguiſh a ranunculus from other families,
which in their outward appearance much
reſemble it.

The different génera of flowers are more
eaſily diſtinguiſhed from each other than, from
their firſt appearance, might be imagined,
though rarely by ſo obvious a character as
this of the ranunculus ; yet, in the ſtudy of
the ſyſtem of vegetables, it will be found that
very minute circumſtances, and ſuch as in
the common obſervation of a flower might be
overlooked, have been made uſe of to mark
not only one family, but every individual of
that family, from each other.

The lady-ſmock (cardámine) is a proper
ſpecimen of a croſs-form flower ; the lung-
wort (pulmonária), of the funnel-form ; the
thyme (thy'mus), of the grinning ; the broom
(ſpártium), of the butterfly. The larger kind

of

of flowers are thofe which fhould be made choice of by the young ftudent for diffection, as their parts are more diftinctly vifible; the crown-imperial (fritilláira imperiális), the poppy (papáver), and the tulip (túlipa), are well fuited to this purpofe, although there are circumftances in each which may perplex a novice in the fcience. The calyx of the poppy falls off immediately when the flower expands; the crown-imperial and the tulip have not any. Linneus efteems only two parts of fructification neceffary to conftitute a flower, in the language of botany, though, perhaps, there might properly be added a third, the *Nectary:* the calyx is the part wanting in the tulip and crown-imperial; but when only one of thefe covers is found, it muft not be inferred to be the corol becaufe it is not green. Although in moft cafes the Corol may be known by the gaynefs of it's colour, or by it's not enclofing the feeds, there are too many exceptions to thefe rules to allow them to be wholly relied on. The petals in paffion-flower (paffiflóra) are green, like the leaves; the corol in Selágo enclofes the feeds. The calyx and corol may, however, be diftinguifhed by the following rule: the ftamens

and

and petals are found to be ranged alternately
in the complete flowers; that is, such as
have both Calyx and Corol of the fourth and
fifth classes of Linnæus's system; hence this
is concluded to be their most natural situa-
ation, while the stamens are placed opposite
to the divisions of the Calyx. Linnæus seems
to consider this as a constant mark; yet he
terms the single cover of many plants of the
sixth class a Corol, in contradiction to this
rule. There is only one cover present in the
crown-imperial, the stamens and petals are
placed alternate; it is therefore a *Corol.*
Although a close observance of this rule would
lead to error in the examination of many of
the beautiful flowers of the sixth class, it will
be expedient for the pupil in botany to follow
Linneus in the term he has given to the only
cover that will be found, and call it the Corol, leaving these small defects of his system
to be corrected by those who, from being acquainted with it's superior merit, are more
desirous to contribute their efforts to render
it perfect, than to expose and cavil at the few
errors which may be discovered in a work of
such superior genius and extensive utility.
The crown-imperial has all it's parts except

the

the calyx; the corol is fix-petalled and belled:
the grace with which the beautiful bell-
flowers are hung round the fummit of a tall,
rich, green ftem, and the elegant appearance
of the tuft of narrow fhining leaves rifing
from the midft of them, with the fmall cavity at the bafe of each petal filled with a
pure cryftalline liquid, render the whole one
of the firft objects of admiration to all who
have a tafte for the natural beauties of a
flower garden. Nor is the outward appearance of this lovely plant alone worthy of admiration; the honey drops contained in the
cavities at the bafe of each petal are objects
of much curiofity, the quantity being fo nicely
adapted to the parts by which it is contained;
as to preferve them always full and apparently
ready to overflow, and yet never to exceed
it's proper limits. The ftamens and piftils of
crown-imperial are very confpicuous; each
particle of duft, when viewed through a microfcope, exhibits the moft perfect form. The
ftyle and ftigma fhould alfo be examined: we
may perceive, with the naked eye, the moifture
at the top of the ftigma, which fits it to
receive the duft of the Anther, and to convey it's effence through the ftyle to the

3   Germe;

Germe; when this Germe becomes a *Pericarp*, or, in other words, when it arrives at maturity, it is a Capfule filled with large flat feeds. There is no peculiar curiofity in the Receptacle of the crown-imperial, nor does there often occur any in the common claffes of flowers. There is a part which may be miftaken in fome flowers for their Calyx; this is what is termed the Bracts, or Floral-leaves; thefe are fituated on the petiole, or flower-ftalk, and often fo near the fructification as to be confounded with the Calyx. Examples of the Bract may be feen in tilia (lime-tree), monárda, paffiflora, paffion-tree; the Bracts may be diftinguifhed from the Calyx by their longer duration; they differ in fize, fhape, and colour, from the other leaves of the plant, but commonly continue as long as they do; whereas the Calyx always withers when the fruit is ripe, if not before. An inftance of this kind of Bract is feen in the beautiful bunch of leaves which rifes among the flowers of crown-imperial, and which has juft now been defcribed. There is a fpecies of fage (falvia) the Bracts of which are beautifully coloured; fometimes they are red, and fometimes of a deep blue. Linneus has made

D great

great ufe of thefe fingularities in determining the fpecies of plants; hence it is neceffary they fhould be well underftood. The Bract is ranked amongft the Fulcra or fupports of plants, which will be made the fubject of the next lecture. The poppy and tulip fhow the ftigma attached to the germe, without the intervention of the ftyle; the germe of poppy with it's ftigma is very beautiful; the ftigma fhuts up the germe, like the lid of a box; when the germe is mature, it is of that fpecies of feed-veffel called a *Capfule*, and opens at the top in feveral places to give paffage to the feeds, which are very numerous. From one head of white poppy 8000 feeds are faid to have been produced in one fummer. This has been afcertained by counting the number of feeds, which would weigh a grain or two, and then by weighing the whole. Seeds of all kinds well repay the trouble of examination, when, viewed through a microfcope, infinite beauty appears in their conftruction, which, from the minute fize of many of them, is loft to the naked eye. The variety that may be found in feeds is very great, both in fize, fhape, and furface, alfo in the veffels which contain, and the fubftance which enclofes

clofes them, before they are ripe. If the difference in the fize of the cocoa-nut feed, and
that of the poppy, be confidered, it will be
obvious, that the fizes muft be very various
between thefe two extremes. The appendage
which nature has given to feeds for the purpofe of their diffemination, frequently is a
great addition to the beauty of their appearance. The feed of common chickweed is a
beautiful microfcopic object, the furface refembling the *Murex* fhell; and a knowledge
of a great variety of feeds may be agreeably
acquired from the elegant coloured engravings
of many different fpecies in Mr. Curtis's
London Flora.

Linneus has named thofe parts of plants,
the chief ufe of which is to ftrengthen and
fupport them, Fulcra, or Props; *fupports* is
the term given them in the tranflation of
the fyftem of vegetables: they are defined to
be, affiftances for the more commodious fupport of the plant. There are *feven* kinds of
Fulcra, or Supports: Petiole, Peduncle, Stipule, Tendril, Pubefcence, Arms, Bract.
Petiole is the foot-ftalk of a leaf, which it
fupports without any flower. Peduncle is the
foot-ftalk of the flower. Petiole is defined to

be

be a prop fupporting the leaf. Peduncle, a prop fupporting the fructification. Stipule is a fcale, or fmall leaf ftationed on each fide of the bafe of the Petioles, or Peduncles, when they firft begin to appear, as may be feen in the Papilionaceous, or butterfly-fhaped flowers. The ftipules of all plants fhould be attended to, as they frequently ferve to diftinguifh one fpecies from another; thofe of the tulip-tree (liriodéndron) are particularly obvious, con-fifting of two large bluifh fcales: within thefe are depofited the infant leaves of the plant, which may be often found fo fmall as to render a microfcope neceffary to the accurate ex-amination of them, when they will be found perfectly formed in every part. By the Sti-pules they are protected and cherifhed until they have acquired fufficient ftrength to fup-port themfelves. The Stipules of the plane-tree (plátanus) add much to the beauty of the tree in fpring, being formed like little ruffs which furround the branches. In peach (amy'gdalus) and bird-cherry (prúnus) the Stipules refemble two very fmall narrow leaves, and are feated at the bafe of the Petiole of the common leaves. The Tendril is a fpecies of Stipule with which every one is acquainted;

thofe

thofe plants are generally furnifhed with this
kind of Stipule, which are not ftrong enough
to fupport themfelves. Vines (vítis) twift
themfelves round other trees by their clafpers
or tendrils, and thus raife themfelves from the
ground. Long poles are placed in our hop-
yards for the fupport of the hop plants (húmulus), which make a very elegant appearance in their moft luxuriant feafon; their
natural place of growth is in hedges, where
they readily find fupporters: all thefe climbing plants are in fome degree injurious to the
tree of which they take hold for fupport, as
they deprive it of that fhare of light and air
to which it has a natural right. There are,
however, fome fpecies of climbers which feem
intended by nature to receive their nourifhment from other plants, as dodder (cufcúta).
The feed of this plant fplits without Cotylédons, fo that the young plant, having no
ftore of nourifhment laid up for it by nature,
feems neceffitated inftantly to find a fofter-
mother, or to perifh; when the feed fplits it
protrudes a fpiral body, which, without
making any attempt to root itfelf in the
earth, afcends the vegetables in its neigh-
bourhood, twifting round them, and abforbing

D 3   it's

it's nourifhment by veffels apparently inferted
into it's fupporters: this muft injure the plants
on which it lives materially. Nor is this the
only way by which it is deftructive to it's
fofter parent; for no fooner does it arrive at
a ftate of ftrength and vigour than it expands
it's branches, and overpowers and fmothers
it's protector. There are but few inftances of
fuch plants as cufcuta in the vegetable kingdom. In moft fituations the injury is fmall,
which the fupporters of the climbing plants
fuftain from the affiftance they afford to their
more feeble brethren, as, generally, climbers
have roots which ftrike into the earth, and
thence draw nourifhment. Some of this
tribe of vegetables are made ufe of at our tables; the tops of hop plants are much fought
after in fpring.

Climbing plants are of fuch quick growth
that their tops are always tender, and, when
rendered mild by boiling, are agreeable food.
The tops of white bryony (bryónia) are faid
to be fweet and pleafant to the tafte. There
is one plant of the parafite kind the hiftory
of which is curious, as it appears to be fo
from choice· it firft vegetates in the earth, and
is fometimes found growing in it; nor has it
any

any want of fupport from it's neighbours, being a ftiff fhort-ftemmed plant; this is the
orobánche major; it grows upon the roots of
other plants, chiefly upon the butterfly-flow-
ered tribe: it has an extremely fmall feed,
which makes it difficult to fhow it's vegetation
by experiment, more particularly as it requires a peculiar foil and fituation for it's culture. Mr. Curtis, in hís London Flora, gives
a plate of it, and fuppofes, that, when the
feed has firft vegetated in the earth, the Radicle fhoots downwards, till it finds a proper
root to attach itfelf to; that then it quits it's
parent earth, and becomes parafitical. In this
ftate it is frequently found upon broom hills,
the roots of the common broom (fpartium
fcopárium) being peculiarly grateful to it;
though, when it contents itfelf with the earth
for it's nutriment, it grows in corn-fields and
on hedge-banks. The fifth kind of Fulcra,
Pubefcence, might, perhaps, have been more
properly denominated a defence than a fupport. This term is applied to every kind of
hairynefs which exifts on plants. If the young
parts of plants be examined by a microfcope,
particularly the young ftalks or ftems, almoft
all of them will be found covered with hairs:

D 4   this

this clothing in their tender ſtate ſeems intended to preſerve them from ſevere winds,
and from the extremes of heat and cold,
which purpoſe it is well adapted to anſwer.
Arms is the general term for thoſe points,
which prevent animals from injuring the
plants; theſe arms conſiſt of Prickles, Thorns,
Forks, and Stings. The ſhrubs and trees which
have Prickles and Thorns for their defence are
grateful food to animals, as gorſe (úlex) and
gooſeberry (ríbes), and would be quickly devoured, if not thus armed. The large hollies in
Needwood Foreſt are armed with thorny leaves
about eight feet high, and have ſmooth leaves
above; which is a curious circumſtance, as it
would ſeem to imply a conſcioufneſs in the
trees, that when their branches were out of
reach of the deer, they had no occaſion for
arms. But though they may thus preſerve
their lower branches from the attacks of the
deer, they cannot defend themſelves from the
depredations of the keepers, who lop their
upper boughs in winter, and ſtrew them on
the ground, and thus furniſh their herds with
a grateful food, when herbage is ſcarce. The
deer peel off the bark from theſe branches
with great dexterity; and this with the
ſmooth

fmooth leaves forms a great part of their
fuftenance in fevere winters. Stings, as in
nettles (urtica), are the pipes of a fmall bag
furnifhed with a venomous fluid; when the
fting, or point, has made the wound in the
finger, which has touched the plant, this fluid
paffes into it, and caufes acute pain. There
are many curious contrivances for the defence
of plants, which may be confidered as arms.
On the leaves of Venus's flytrap (dionæa muf-
cipula) there is a wonderful contrivance to
prevent the depredations of infects; the leaves
are armed with long teeth, and lie fpread upon
the ground round the flower-ftem, and are fo
irritable, that, when an infect creeps upon them,
they fold up, and pierce or crufh it to death.
We have a plant of our own country, which, in
it's curious mechanifm, greatly refembles the
fo much celebrated flytrap; this is the fun-
dew (drófera*): it's round flat leaves are thickly
befet with hairs, both on their upper furface
and on the margin; each of thefe hairs is
crowned with a little purple globule, which
in the funfhine exudes a pellucid drop of
mucilage, and gives the whole plant a beau-

* See Plate the Third.

tiful

tiful appearance. Thefe hairs with their vifcous juice entangle the flies, which attempt to plunder the leaves, fo completely, that, when once enclofed by them, it is not poffible they fhould efcape. It is alfo fuppofed, that the leaves of the drófera poffefs a power of folding themfelves upon the infect, that they would deftroy, in a manner fimilar to thofe of the flytrap. This elegant little plant grows commonly upon marfhes, and upon wet parts of heaths and on ditch banks; in thefe fituations they are not difficult to difcover, as they form a little red patch, which immediately attracts the eye. There is alfo a vifcous juice furrounding the ftems of fome plants, which effectually defends them from the depredations of infects, as they cannot extricate themfelves from this glutinous material, if, by an attempt to fettle upon the ftalks, they become entangled by it; from this circumftance a fpecies of Siléne has obtained the common name of catch-fly. There are many more extraordinary arts, which nature has ufed to preferve the vegetable kingdom from it's numerous enemies of the animal creation. This curious and interefting part of the fubject of botany muft, however,

be

be referved for proficients in the fcience, as it more properly belongs to the philofophical part of that agreeable ftudy. The Bract, or floral loaf, has been explained in the laft lecture. There is another kind of flower-ftalk befide the peduncle, which is termed *Scape*. The Scape is that kind of flower-ftem which raifes the fructification without the leaves; it is a naked ftalk proceeding immediately from the root, and terminated by the flowers. Hyacinth (hyacinthus), lily of the valley (convallária), and áloe, are examples of the Scape. The fmall ftalk belonging to each flower is termed a Peduncle. An acquaintance with the different kinds of flower-ftalks is effential to an accurate knowledge of the various modes of Inflorefcence, a term which fignifies the various manners in which flowers are joined to their Peduncles. There are feven different modes of Inflorefcence, diftinguifhed by the following terms: Verticil, Head, Spike, Corymbe, Thyrfe, Raceme, Panicle. The Verticil is that kind of Inflorefcence where many flowers furround the ftem like a ring, or ruff, the individual flowers ftanding upon very fhort peduncles, deadnettle (lámium), and lavender (lavandula),

bear

bear their flowers in a Verticil, or Whorl.
Head has many flowers collected into a globe
on the fummit of the common ftalk, fometimes with, and fometimes without, diftinct
Peduncles. Clover and globe amaranthus
(trifólium and gomphréna) fhow this kind of
Inflorefcence; it is diftinguifhed into various
kinds by it's fhape and other circumftances.
Sweet William (diánthus barbatus) has it's
flowers in that fpecies of head which is called
a Fafcicle, though it feems that the mode, in
which the flowers of fweet william are put
together, would place it more properly under
the term Cyme than Head. The Spike has
it's flowers placed alternately round a common fimple peduncle, without any partial
ones, which is called being feffile, or fitting
clofe on the ftem. Many of the graffes have
their flowers in Spikes: a Spike is called
one-ranked, or a fingle-rowed fpike, when
the flowers are all turned one way following
each other; a double-rowed fpike, or two-
ranked, when the flowers ftand pointing two
ways, as in darnel (lólium). The Spike,
like the Head, is diftinguifhed into various
kinds by it's fhape, and other varieties. The
Corymbe is formed by the partial peduncles
                                    produced

produced along the common ftalk on both
fides, which, though of unequal lengths, rife
to the fame height, fo as to form a flat and
even furface at top. Spiræa opulifolia, and
candy-tuft (ibéris), alfo are examples of the
Corymbe. The earth-nut and parfley refemble
the Corymbe in their manner of flowering:
there is, however, this diftinction, the flowers
which form the general bunch of parfley (ápium)
and earth-nut (búnium), which is called an
umbel, all grow from the fame centre; whereas
thofe of the Corymbe grow from different
parts of the common flower-ftalk. The Thyrfe
is the mode of Inflorefcence we have now to
confider. The flower of lilac (fyrínga), and
of butter-bur (tuffilágo), are examples of the
Thyrfe. Linneus calls it a panicle condenfed
into an egged form ; the lower peduncles,
which are longer, extend horizontally, or
crofs-way; the upper, which are fhorter,
mount vertically, or in a perpendicular direction. The raceme has it's flowers placed on
fhort partial peduncles, proceeding like little
lateral branches from and along the common
peduncle; the raceme refembles a fpike in having the flowers placed along the common peduncle; but differs from that mode of inflorefcence

cence in having partial peduncles; it alſo differs
from the corymbe in the ſhortneſs and equal
length of it's peduncles, not forming a regular
ſurface at top. The vine (vítis) and the
currant (ríbes) bear their flowers in Racemes.
The Panicle. has it's flowers diſperſed upon
peduncles, variouſly ſubdivided, and is a
branching diffuſed ſpike, compoſed of a number of ſmall ſpikes, that are attached along a
common peduncle. Oats (avéna) have their
flowers in Panicles.

We have now gone through the various
terms given by Linneus for the manner in
which flowers are placed upon their peduncles,
all of which are ranked under the term Infloreſcence, and ſhould be carefully impreſſed
upon the memory. Flowers are alſo ſometimes found growing on the leaves, as in
the genus of Rúſcus. Dr. Thunberg takes
notice of this ſingular kind of infloreſcence in
his account of Japan, having ſeen it in the
Oſy'ris Japónica, and calls it a moſt rare circumſtance in nature. From it's rare occurrence,
probably, Linneus has not thought it neceſſary to diſtinguiſh this mode of infloreſcence
by any particular term, though in the rúſcus,
where it occurs, he calls it leaf-bearing. The

3   umbel,

umbel, which has been before explained, the
cyme, and the ſpadix, he has ranked under
the general term Receptacle. The cyme and
umbel are much alike, both having a number
of ſlender peduncles growing from one common centre, which riſe to the ſame height;
they differ, however, in the cyme having it's
partial peduncles diſperſed along the ſtalk
without any regular order. Elder (ſambúcus)
and lauruſtinus (vibúrnum) are ſpecimens of
the cyme. The term Spadix is uſed to ex-
preſs every flower-ſtalk that is protruded
from a ſpathe or ſheath; the family of palms
have their flowers in a ſpadix, which is
branched. The ſpadix of all other plants is
ſimple. There is yet another term, which
Linneus makes uſe of, which is Rachis; this
means only the ſtem, on which the flowers
grow that form a ſpike. He defines the Rachis
to be a thread-form receptacle, connecting
the florets longitudinally into a ſpike. There
may appear much difficulty in the attainment of an acquaintance with ſo great a
variety of terms which convey no preciſe
ideas; an attentive conſideration of them,
with a compariſon of the definitions of the
different kinds of Fulcra and modes of Infloreſcence,

refcence, with the drawings of them in Plate the Third, will, however, render the tafk by no means a hard one. Botany has been reckoned a dry ftudy of names and terms; and this view of the fcience has deterred numbers from attempting to acquire a knowledge of it. This is by no means peculiarly the cafe; every fcience has a language appropriate to itfelf; every language has a grammar: thefe difficulties muft be furmounted before the fcience or language can afford entertainment. In Botany, however, inftruction and amufement may be united, if, as the pupil proceeds, he examines and compares the different parts of flowers with the terms appropriated to them. By this means the beauties of nature will open to his view, and he will in the very commencement of his ftudies obtain a glimpfe of that wonderful order and mechanifm, which are to be found in the vegetable creation, and which render botanical purfuits fo completely interefting.

EXPLANATION

EXPLANATION OF PLATE III. PART I.

Fig. 1. A Seed of Cucumber, *a*, before it is put into the ground. *b*, Beginning to germinate. *c, c.* The Cotylédons expanded. *d*, The Plume. *e*, The Radicle.

Fig. 2. The Seeds of Geranium, to ſhow the manner in which they are diſperſed. *f*, The Awns by which they are attached to the Piſtil.

Fig. 3. The common Receptacle of a Compound Flower.

Fig. 4, and 5. Different ſhaped Florets of Compound Flowers.

Fig. 6. The Wheel-form Corol of Verónica, to ſhow the narrow diviſion.

Fig. 7. A Petal of common Crow-foot. *g.* The Neétary.

Fig. 8. Shows a Tendril, *h.* Stipules, *i.* Glands, *k.*

Fig. 9. A Verticil.

Fig. 10. Head.

Fig. 11. A Spike.

Fig. 12. A Corymbe.

Fig. 13. A Thyrſe.

Fig. 14. A Raceme.

Fig. 15. A Panicle.

Fig. 16. Leaf-bearing.

Fig. 17. An Umbel.

Fig. 18. A Cyme.

Fig. 19. A Braét, of Lime Tree (Tília Európæa) with the Capſules mature.

Fig. 20. A Plant of Dróſera, Sun-dew.

London, Published May 1.st 1797, by J. Johnson, St. Pauls Church Yard.

LECTURE III.

The firſt eighteen Claſſes, with their Orders, explained.

A PREVIOUS knowledge being acquired of the ſeven parts of *Fructification*, with all their variations; the different kinds of *Fulcra*, and modes of Infloreſcence, being well underſtood; the pupil may proceed to the Claſſes.

A Claſs is the firſt and higheſt diviſion of every ſyſtem. It may be compared to a dictionary, in which all the words having the ſame initial letter are arranged together, every word may be compared to a genus; the claſſic character is conſtituted from a ſingle circumſtance, as the words are arranged by a ſingle letter; this one circumſtance muſt be poſſeſſed alike by every plant admitted into the Claſs, how different ſoever they may be in other reſpects. This ſingle character is arbitrary, and has been taken from various parts of the fructification by different authors; ſome have choſen the petals, others the fruit; Linneus has made choice of the ſtamens, and

E 2   on

on their number and fituation has founded
his claffes; he makes the excellence of the
claffic character to confift in it's greater or
lefs approximation to the natural one. The
claffes called natural are thofe which contain
plants agreeing in a variety of circumftances,
fuch as habit, manner of growth, ufes, and
fenfible qualities. The graffes are a natural
clafs; the compound, the pea-bloom, the
crofs-form, the umbelled, and the verticilled
plants, are natural claffes; fo are the ferns.
Though fome of Linneus's claffes are natural,
moft of them are artificial; this, however, is,
perhaps, of little confequence; his fyftem has
opened to our view a diftinct knowledge of
every plant that grows; it has given us a
clear and ready method of referring an unknown plant, 1ft, to it's Clafs; 2d, to it's
Order; 3d, to it's Genus; 4th, to it's Species;
and 5th, to it's Varieties. Before we had this
ingenious fyftem to guide us to a knowledge
of the vegetable kingdom, all was confufion.
Much acutenefs had been difplayed in the
inveftigation of plants; but the labours of
many ingenious men were rendered of little
ufe from want of arrangement; they claffed
plants together which had fcarcely any affinity,

from

from a fancied refemblance in imaginary virtues. Much ufeful knowledge has been loft to
the world, almoft all the medicines, and many
of the arts of the ancients, we are now ignorant of, from their deficiency in the knowledge of Botany.

But, notwithftanding this deficiency in arrangement, we muft not overlook the merits
of the old writers on this agreeable fcience;
to our own countrymen, Dr. Grew and Gerrard, we ought to be particularly grateful.
Dr. Grew made his inveftigations with an eye
fo penetrating and accurate, that much information may be found in his book on the
anatomy of plants, particularly in the philofophical part of Botany; befides, it is pleafing
to obferve the coincidence of his opinions
with thofe of Linneus, in regard to the ufe
of the parts of fructification. Gerrard's defcriptions are full and ftrong, and his language amufing; but, from want of arrangement, the ftudent is bewildered, when he
looks for a plant in his Herbal. The various
fyftems of modern botanifts have defervedly
had their partifans; but it now feems generally allowed, that the works of Linneus are
beft calculated to enable us to attain a know-

E 3   ledge

ledge of botany. He has divided the vegetable kingdom into twenty-four Claffes; the firft ten Claffes include the plants in the flowers of which both ftamens and piftils are found, and in which the ftamens, when arrived at maturity, are neither united nor unequal in height. Thefe Claffes are therefore diftinguifhed from each other fimply by the number of ftamens in each flower, and may be known upon the firft view by their numbers, as expreffed by the words prefixed to the Claffes: the firft Clafs is known by the name of Monandria, which fignifies one-male, or one-ftamen, the ftamens being the part of fructification, which Linneus calls the male; fo that the numerical word joined to the word ándria forms the titles of the firft thirteen Claffes; an attention to which circumftance will make the tafk of committing them to the memory by no means difficult. An enumeration of the titles of the firft thirteen Claffes may be of ufe. Monándria, one-ftamen; diándria, two-ftamens; triándria, three-ftamens; tetrándria, four-ftamens; pentándria, five ftamens; hexándria, fix-ftamens; heptándria, feven-ftamens; octándria, eight-ftamens; enneándria, nine-ftamens; decándria, ten-ftamens; dodecándria, twelve-

twelve-ftamens; icofándria, twenty-ftamens; polyándria, many ftamens.

The pupil fhould render himfelf familiar with the titles of the Claffes compounded by Linneus, equally with thofe which are formed in his own tongue; for although, in moft elementary works intended for the ufe of the englifh ftudent of botany, an attempt has been made to bring englifh terms, and names of plants, into ufe in preference to thofe employed by Linneus, fuch language cannot anfwer the purpofes of a general botanift; the pupil of thefe authors cannot converfe with one of the Linnean fchool. In the tranflated works of Linneus he will learn a language which will enable him to communicate with botanifts of all nations, and to underftand any botanical defcriptions of plants that he may meet with. They who have not induftry fufficient to ftudy thofe books will learn the fcience in but a fuperficial manner from any. The complaint, that the tranflated works of Linneus are hard, arifes from not knowing how to ftudy them. The method adopted in thefe Lectures may, I hope, enable my pupils to become proficients in this agreeable feience with as little difficulty, and more amufement,

E 4   than

than from any of the various circuitous ways
which have been made use of to level the
subject to the capacity of ladies. Twenty
years ago an englifh botanift, defirous to be
acquainted with the fcience, might with reafon complain of the hardnefs of the ftudy;
but at this enlightened period knowledge is
fo widely diffufed, that there are few fituations
where books, with plates of explanation, are
not to be met with, or fome friend to be had
accefs to, who is both able and willing to
elucidate any obfcure expreffion which may
occur.

But to proceed with the Claffes, the ten
firft of which are reprefented in Plate the
Fourth, and are diftinguifhed by the number
of their ftamens only; the eleventh clafs is
called dodecándria, which fignifies twelve-
ftamens. The reafon of paffing from ten to
twelve is, that the number eleven has not
been found fufficiently conftant in any
flowers to form a Clafs. In the genus refeda
eleven ftamens are fometimes found, but
oftener more; yet they never exceed fifteen.
The effential character of the eleventh Clafs
depends on the flowers belonging to it having
fewer than eleven ftamens, and not exceeding
nineteen;

nineteen : added to this may be, that in this Clafs the ftamens are fixed to the receptacle; whereas in the next, which has the title of twenty-ftamens, icofandria, though not more determined in point of number than the preceding one, they are attached to other parts of the fructification : their pofition it is alfo neceffary to attend to in the thirteenth clafs; fo that if we regarded only the titles of thefe three claffes, we fhould find ourfelves much confufed. This is certainly a material defect in the fyftem, which cannot be accounted for in a fatisfactory manner. Linneus was evidently aware of the imperfection in the titles of thefe Claffes, and has guarded againft the inconvenience which would arife from the firft character expreffive of a decided number of ftamens, by adding in the Key to his fyftem the fituation of their growth, by which circumftance alone we can diftinguifh thefe three claffes one from the other. The twelfth clafs, icofandria, has generally twenty ftamens, often more, which are inferted on the calyx; there are alfo other more obvious characteriftic marks, which may ferve to diftinguifh this twelfth clafs from the following one, and which fhould be attended to, as this contains

moft

moft of the wholefome fruits, and the thirteenth chiefly confifts of fuch plants as are
poifonous; and it is curious to remark how
juftly the infertion of the ftamens into the
calyx may be relied on as an indication of a
fruit free from noxious qualities. In the
Prunus genus there are fome fpecies, as the
padus and lauro-cerafus, in which every part,
except their pulpy fruit, is poifonous; and of
that we may eat with fafety. This-mark is
alfo worth attending to in the plants of other
claffes. In the clafs Pentándria Monogynia
there are many fruits, the juices of which are
highly deleterious; but in Ríbes (currant and
goofeberry) we find a wholefome and grateful fruit, indicated by the circumftance of the
infertion of the ftamens into the calyx. This
charaƈteriftic diftinƈtion of the clafs Icofándria
is alfo vifible when the fruits are ripe, their
calyx frequently remaining like a little crown
on their top, and, while in a frefh ftate, a
fkilful botanift may diftinguifh the infertion
of the ftamens on the inner part of it's divifions. The flowers of the twelfth clafs, Icofándria, have a hollow calyx of one leaf, the
corol faftened by it's claws to the infide of
the calyx, and, as was before obferved, the

                                        ftamens

ftamens placed on the infide of the calyx or corol. The thirteenth clafs, many *ftamens*, Polyándria, has it's ftamens inferted on the receptacle; their number being from twenty to one thoufand in the fame flower. This clafs is the laft of the numerical ones, or, more properly, of thofe which have numerical titles, it having been fhown that the characters of the three laft claffes depend nearly as much on the fituation of the ftamens, as thofe which are yet to be confidered. The firft thirteen Claffes, with their Orders, fhould be well underftood, before thofe which are more complicated are entered upon.

The Claffes are all divided into what are termed *Orders*; thefe fubdivifions of the firft thirteen Claffes are founded on the number of piftils, or on that part of fructification which Linneus calls the female. If a flower contains one of thefe females or piftils, it is of the firft order; if it contains two, of the fecond; and fo on to any number that it may contain. The Linnean term for the orders is formed from the Greek word, which fignifies a female, joined to another word expreffive of the number; fo that, as Monándria fignifies one-male or ftámen, Mo-

nogynia

nogynia means one female or piftil; Digy'nia
fignifies two piftils, which refers the plant to
the fecond order; Trigynia fignifies three;
and in the fame manner the terms proceed to
Polygynia, or many piftils.

The prefence of the female part of fructification, or the piftils, is equally neceffary with
that of the male, or the ftamens, to conftitute
a flower belonging to the firft thirteen Claffes;
and it muft alfo be remembered that the ftamens, when at maturity, muft be of an equal
height. The effential character of the clafs
Dodecándria, or the eleventh clafs, may be
feen in the flowers of reféda odorata, mignonette; the ftamens will be found to be not
lefs in number than eleven, nor to exceed
nineteen, and to be fixed on the receptacle.
The diftinction between the claffes Icofándria
and Polyándria, twenty ftamens and many
ftamens, may be well feen in the bloom of
apple, and in the flowers of the common
crow-foot, ranunculus arvénfis; in the apple
bloffom there are generally twenty ftamens,
often more, inferted upon the calyx, which
is of one leaf, with the claws of the corol
faftened on the infide of it; in the crow-foot
the ftamens are moft numerous, and all attached

tached to the receptacle. The clafs Didynámia, two-powers, or the fourteenth clafs,
is diftinguifhed by the flowers which are contained in it having four ftamens, two of them
being longer than the other two; hence it is
called the clafs of two powers. The grinning
and gaping flowers belong to this clafs. There
are, however, two fuch diftinct natural affem-
blages of plants contained in it, that it would
have been difficult to have brought them together from their affinity in any one circumftance, but that under which Linneus has arranged them, viz. the curious pofition of their
ftamens. This clafs contains two orders,
which are ftrongly marked; the firft gymno-
fpermia, or that in which the flowers have
their feeds naked, being contained in the
bottom of the calyx; and the fecond order,
angiofpermia, having the feeds covered or
contained in a pericarp. The whole appearance of the flowers belonging to thefe two
orders is perfectly different: what can be more
fo than the fox-glove (digitális), and lavender
(lavandula), or thyme (thymus)? Yet the
crofs-form growth of the anthers, with the
unequal pofition of the ftamens, may be found
in them all. The next clafs, Tetradynamia,
four-

four-powers, or the fifteenth clafs, has fix ſtamens, and is called the clafs of four-powers:
theſe fix ſtamens not being of an equal
height, four being taller, and the two lower
growing oppoſite to each other. This clafs
contains the crofs-form flowers, and is a really
natural clafs. Linneus has admitted only one
genus into it which can be at all objected
againſt, that is the genus cleóme, in many
ſpecies of which there are more than fix
ſtamens, and theſe not in the regular proportion of length, which gives the name of
four powers to the clafs, fo that it ſeems that
the family of cleóme has no right to be admitted into it, unleſs the affinity of it's nectaries to thoſe of the crofs-form flowers may
be allowed a ſufficient title. This clafs is divided into two orders, which are diſtinguiſhed
by the form of their pericarps, or feed-veſſels;
the firſt order having it's feed-veſſels of the
Silicle kind, the fecond of the Silique; the
Silicle being furniſhed with a ſtyle, often the
length of itſelf, the Silique with a ſtyle ſcarcely
viſible. The ſilicle of honeſty, when mature,
is a great ornament to the plant; from its
ſhining appearance, like white fattin, it has
received it's botanical name of lunária, or
moonwort.

moonwort. There is a good deal of variety in the forms of the filicle kind of feed-veffel; that of lunária is nearly round; there are others which are oval: the fmall filicle of fhepherd's purfe (thláfpi) is triangular, and notched at the top, and refembles a little heart; the circumftance of being notched or plain makes two divifions of the filicle order, and thence renders the inveftigation of the génera belonging to it a lefs difficult tafk. The feedveffel of lady fmock (cardamíne) is a filique, and alfo that of radifh (ráphanus). Some of thefe filiques form very pretty fkeletons, in the manner of thofe holly leaves which have lain on the ground and been expofed to the weather in winter. The fixteenth clafs, Monadélphia, or one-brother hood, is fo called from the flowers belonging to it having all their ftamens united at the bafe into one company, furrounding the piftils. The ftamens and piftils in the flowers of the fixteenth clafs form a beautiful part of the fructification; they ftand like a little pillar in the centre of the flowers, from which circumftance Linneus, in his Natural Orders, has named thefe flowers column-bearing. The anthers have a marked character, which contributes to their

3   ornament,

ornament, being fhaped like a fmall kidney, and attached to the filaments by their middle in fo flight a manner, that they appear rather to lie upon than to be fixed to them. The piftils are enclofed by the ftamens, till they begin to advance towards maturity, when they burft forth, and form an elegant taffel, a little above the furrounding anthers: in the china rofe (hibífcus) this taffel is particularly beautiful; the rich crimfon piftil rifes rather higher than ufual above the golden anthers, which encircle it, and dividing into five filaments at top bends down it's round ftigmas amongft them; thefe ftigmas, at the period of maturity, having the appearance of the richeft crimfon velvet fpangled with gold. The double hibifcus is that which is generally cultivated; but it is greatly inferior in beauty to the fingle, as, from the multiplication of it's petals, the other elegant parts of the fructification are excluded. As the fixteenth clafs is founded on the fituation of the ftamens, fo are the orders on their number, beginning with the number three, and ending with that of eleven. The clafs Diadélphia, or two-brotherhoods, the feventeenth clafs, is perfectly natural, and the ftructure of the corol fo

remarkable,

remarkable, that the outer habits of it's flowers
are fufficient to diftinguifh them from all
others; but, according to the Linnean fyftem,
it is neceffary to have recourfe to the fituation of
the ftamens, which he defcribes as being united
into two fets; this claffic character is, however,
to be traced with difficulty, for what is termed
one of the fets, confifts of a fingle filament;
and even this obfcure mark does not exift in
all the genera; indeed, fo many are deftitute
of it, that Linneus has, on this failure, founded
one of the fubdivifions of the fourth order.
He has, however, efteemed it of fuch effential
confequence, that he has excluded from the
clafs the genus Sophóra, which has all the
characters of the Diádelphia tribe, except that
of the united filaments; and on this fingle
deficiency he has feparated it from it's natural
tribe, and placed it according to it's number
of ftamens, which is ten, in the clafs Decándria, with the flowers to which it has no
affinity in any other parts of the fructification.
The orders, or fcondary divifions of the feventeenth clafs, are founded upon the number of
ftamens, without any reference to their union;
the fingular ftructure of the corol having made
it neceffary to diftinguifh each feparate part by

F   a name

a name peculiar to itfelf: the broad fpreading
petal at the back of the corol is called the
Banner; the fide petals, the Wings; and the
two petals, by which the ftamens are enclofed,
are termed the Keel, from the refemblance of
their form to the keel of a boat. The fhape,
and other circumftances attending thefe different parts, are found of ufe in diftinguifhing the genera of this clafs from each other;
but the calyx is of moft fervice in this important office; it is to this clafs of plants
that the legume feed-veffel belongs. The
Legume is diftinguifhed from the Silicle
and Silique by it's feeds being fixed alternately on each fide the edges. The eigh-
teenth clafs is called Polyadélphia, or many-
brotherhoods, the flowers contained in it having their ftamens united into diftinct fets.
St. John's wort (hypéricum) fhows the difpofition of the ftamens very plainly in that
genus; they may, with very little attention,
be taken off in fmall bunches: the orders of
this clafs depend on the number of ftamens,
or, more properly, on the number of anthers in each flower, as fome of the génera
have five anthers on each filament: indeed,
this is a circumftance which ought always
                                        to

to be attended to, the ANTHERS and STIGMAS being the effential parts of the STAMENS and PISTILS. If they are prefent, it is fufficient to place the flower, they belong to, in the clafs or order to which their number refers it.

LECTURE

LECTURE IV.

Examination of Flowers belonging to different Glaſſes.
The·Claſſes 19, 20, 21; *and* 22, *explained.*

As a means to imprefs the knowledge
which has been acquired upon the minds of
my pupils, and in order to render their ſtudies
more amuſing, I recommend to them to attempt to refer ſome plants of ſimple conſtruc-
tion to their claſſes and orders. The young
botaniſt is frequently diſcouraged in his early
endeavours of this kind by the flowers on
which he fixes for his experiments; the whole
tribe of graſſes ſhould be avoided, as they require a peculiar method of ſtudy, and conſider-
able proficiency in the knowledge of botany,
to render them eaſy of acceſs. The ſtate of
the flower, when examined, is alfo an important circumſtance ; the beſt time to examine
the number of ſtamens is immediately before
the corol expands; after the anthers are mature it is difficult, in many flowers, to diſtinguiſh their number. The hippuris vulgáris,
mare's

mare's tail, from the frequency with which
it prefents itfelf to the eye of the young botanift, generally attracts his attention as an
object of inveftigation, and, from the fimplicity of it's conftruction, feems a proper one
for that purpofe, fo far as refpects the characters of it's clafs and order; it has neither
calyx, corol, nor feed-veffel, and thofe parts
moft effential to fructification few as poffible,
there being only one ftamen, one piftil, and
one perfect feed; hence eafily referred to the
firft clafs, Monándria, and the firft order Monogynia: yet fome difficulty is liable to occur
from the mode of inflorefcence, or pofition in
which the fructification is placed upon the
flower-ftalk. A number of florets, containing each a ftamen and piftil fixed at the bafe
of a fmall-pointed leaf, grow round the ftem
in a whorl, and have, to an inexperienced
eye, the appearance of forming only one
flower, though, on accurate examination,
each fmall floret will be found perfect in itfelf, poffeffed of thofe parts which are fufficient to conftitute a fingle flower.

Cánna, flowering-reed, may be more readily
referred to the clafs one ftamen, and order one
piftil, as there are not any difficulties

attending

attending it's mode of inflorefcence. The verónica, common fpeedwell, belongs to the clafs Diándria and order Monogynia. Moft of the graffes may be found in Triándria, three ftamens, but are of a ftructure too difficult for the inveftigation of the young botanift. Crocus is a good fpecimen of the clafs Triándria, but not fo eafily referred by it's characters to the order Monogynia, the deep divifions of the ftigma giving the appearance of three piftils; if, however, the parts of fructification are feparated, to do which the root muft be taken out of the ground, one very long piftil within the tube of the corol will be found. The common plaintain (plantágo) may be referred to the clafs Tetrándria and order Monogynia, four ftamens, one piftil, without much difficulty, if examined before the anthers are arrived at maturity. Several flowers of the fame kind fhould be collected at their different periods of growth; and it muft be remembered, that the four ftamens muft be of equal heights to give the flower a place in the clafs Tetrándria. In the flowers of plaintain the anthers are placed upon very long flender filaments, which, previous to the maturity of the anthers, lie clofely doubled

down

down within the corol to preferve them from injury until they are ready for expanfion. In this ftate it is curious to obferve the unfolding of the filaments, if touched flightly with a fine needle. It is not eafy, in the flowers of the umbel-bearing plants, to find the ftamens in a proper ftate for inveftigation; they alfo differ in number, in which cafe the flower, which terminates the umbel, is to be examined, and, according to the number of ftamens contained in that, is to be claffed. The difficulty of variety in the number of ftamens in the fame fpecies too frequently occurs in the flowers of the clafs Pentándria, and is a perplexing circumftance to young botanifts; but as nature commonly preferves a certain proportion through all the parts of the fame work, the clafs to which a flower belongs may generally be difcovered by attending to the numbers of the other parts of fructification. Should a flower be found which has it's calyx divided into five parts, and it's corol confifting of five petals, though it's ftamens fhould exceed or fall fhort of the number five, it may be concluded, that it belongs to the fifth clafs: and if a few more flowers of the fame fpecies, or even of the fame plant, be

examined,

examined, it will be feen that five ftamens
are the moft conftant number belonging to
fuch flowers; and they may be referred to the
clafs Pentándria without hefitation. The umbelled plants are improper fubjects to begin
with from the minutenefs of their parts of
fructification. The larger forts of flowers,
and thofe of the moft fimple conftruction,
fhould be made choice of, and when they,
with their claffes and orders, are well underftood, the pupil may proceed to more complicated kinds; the honeyfuckle (lonicéra)
and lungwort (pulmonária) are fimple flowers
of the clafs Pentándria and order Monogynia,
five ftamens and one piftil. The fnow-drop
(galánthus), horfe-chefnut (éfculus), and me-
zéreon (daphne), are fpecimens of the claffes
Hexándria, fix ftamens, Heptándria, feven
ftamens, Octándria, eight ftamens, and of
their firft orders, Monogynia, one piftil. The
clafs of nine ftamens, Enneándria, contains
only fix génera. There is but one britifh fpecies known which belongs to this clafs, that
is the bútomus, or flowering rufh, and this is
not to be commonly met with. The wood-
forrel (óxalis) is an elegant fpecimen of the
clafs Decándria, ten ftamens, and the order
Pentagynia,

Pentagynia, five piftils. But there are fome plants placed in this clafs which generally form a ftumbling-block to the young botanift; an inftance of this is found in fome of the fpecies of the family of Lychnis. By a ftrict obfervance of Linneus's rules the lychnis dioica, or two houfe, fhould not be placed in the tenth clafs, as the characteriftic mark of the clafs Decándria requires the prefence of both ftamens and piftils in the fame flower: however, he has himfelf placed it there, being found to agree with the reft of it's family in every particular but that of it's ftamens and piftils being on the fame plant; rather than feparate it from them, he has taken this circumftance for it's fpecific character. This, and a few more inftances of the fame kind, may certainly be confidered as defects of the fyftem; but the inconvenience that might arife from fuch a violation of the general rule, by which the claffes are characterized, is obviated, as much as can be, by being noted whenever fuch contradiction occurs. The ly'thrum (willow-herb) belongs to clafs Dodecándria, twelve males, and is liable to vary in it's number of ftamens, which fhows the neceffity of examining many flowers of

the

the fame genus : however, as the claffic character is not derived folely from the number
of ftamens, fuch variations may be of lefs
confequence. The hawthorn (cratægus) and
pheafant's eye (adónis) exhibit marks of the
claffes Icofándria, twenty males, and Polyándria, many males, the hawthorn having it's
ftamens fixed to the calyx, and thofe of the
adónis being placed on the receptacle. In
the clafs Didynámia, two-powers, Tetradynámia, four-powers, and Monadelphia, one-
brotherhood, the orders or fubdivifions, no
longer depending on the number of piftils,
will require fome farther explanation. In the
fourteenth clafs, two-powers, the génera are
divided into two orders, the firft diftinguifhed
by the feeds being placed within the calyx
without any other covering; the fecond by
the feeds being contained by a pericarp, or
feed-veffel: from thefe different circumftances
the orders derive their names of gymnofper-
mia, feed-naked, and angiofpérmia, feed-
covered. The dead-nettle (lámium) and
fnap-dragon (antirrhinum) are good fpecimens of both orders, and alfo of the clafs
two-powers. The orders of the fifteenth
clafs, Tetradynámia, four-powers, are marked

by

by the form of their feed-veffels; the whitlow-
grafs (drába) is a fpecimen of the firft order;
it's feed-veffel being a Silicle refers it to that
divifion. The feed of purple rocket (héfperis)
being contained in a filique, that genus belongs to the fecond order. We find in the
clafs Tetradynámia many of our efculent vegetables; fome of which, as the water-crefs
(fify'mbrium) and muftard (finápis), are ufed
without having gone through the procefs of
cookery; others are rendered mild by boiling,
as cabbage, turnep, brocoli, cauliflower, and
fome others, all of which are the produce of
cultivation from one genus, Bráffica. The
change produced in vegetables by the art of
gardening is a part of the fubject of botany
highly curious and amufing.

The flowers of the three claffes, Monadélphia, Diadélphia, and Polyadélphia, one brotherhood, two brotherhoods, and many
brotherhoods, are now to be confidered.
The characters of thefe claffes are ftrongly
marked: the geránium and mallow are fpecimens of the Monadélphia clafs; in attempting to take off the ftamens, that union of the
filaments from whence the name of One
Brotherhood is derived, may be diftinctly
feen;

feen; and though apparently feparated at the top they will be found firmly united at the bafe. The orders are characterifed from the number of ftamens found in each flower; the geranium and mallow, having many ftamens, are arranged in the order Polyándria. The form of the papilionaceous, or butterfly, tribe of plants is fo evidently different from that of all others, that no additional mark is re-quifite to diftinguifh them; but in referring thefe flowers to the claffes eftablifhed by Lin-neus, the fyftematic character of Diadélphia, two brotherhoods, muft be examined: this he has made to depend upon the union of the ftamens into two fets, which would lead the botanical ftudent to expect a more equal divifion of the filaments than does in reality exift; the pea (pifum), having a large flower, will give a juft idea of the true pofition of the ftamens; thefe are ten in number, nine of which are feparated from the tenth, and clofely united at the bafe. On this feparation of the tenth filament Linneus has founded his claffical character, not, however, unap-prifed of it's deficiency, as in feveral génera he has made the connexion of all the ftamens the mark by which he collected them under a

fubdivifion

fubdivifion of one of his orders which derive their character from the number of ftamens. In common broom (fpártium fcopárium) the ten filaments are all united; they, however, might, perhaps, with more propriety, be termed two fets than thofe of the pea, five of the ftamens obvioufly rifing a quarter of an inch above the other five. There is a curious circumftance refpecting thefe flowers which is worth attending to: the upper fet of males, or ftamens, does not arrive at maturity fo foon as the lower; and the ftigma, or head of the female, is produced amongft the upper or immature fet; but as foon as the piftil grows tall enough to burft open the keel-leaf, or hood of the flower, it bends itfelf round in an inftant like a French horn, and inferts it's head, or ftigma, amongft the lower or mature fet of ftamens, as may be feen by touching the keel-leaf; the piftil continues to grow in length, and in a few days arrives again amongft the upper fet by the time they become mature. This wonderful fact we owe to the accurate refearch of the much-lamented author of the Botanic Garden, to whom the world is indebted for an ex-tenfive variety of knowledge, both amufing and

and ufeful, and from which benefit will be
derived to mankind to the lateft ages.

In fome génera belonging to the clafs Polyadélphia the charaéter of many brotherhoods
is clearly defined, in others it is lefs obvious;
in the genus Hypéricum, St. John's-wort, it
is eafy to take off the ftamens in diftinét little
bunches. In the orange, lemon, and citron,
all of the genus Citrus, the appearance of the
ftamens differs fo much from that of the hypéricum that a young botanift would not
fuppofe them to be of the fame clafs. However, on inveftigation, the ftamens will be
found feparated into fmall bunches, fo as to
entitle the family to a place among the many
brotherhoods.

The moft intricate clafs in the whole fyftem muft now be confidered: the curious
and beautiful conftruétion of the flowers contained in it will, however, amply repay the
labours of the ftudent. The clafs Syngenéfia,
confederate males, or united anthers, is
founded on the very peculiar fituation of the
anthers, which are joined together in the
form of a cylinder, while the filaments remain
feparate. A flight preffure at the top of this
cylinder of anthers caufes the filaments to
bend

bend down, and diſtinctly ſhews their want
of union: the number of ſtamens ſo united
is five; they form a ring round the piſtil, which
riſes in the midſt of them, and ſeems conſcious
of the homage ſhe is receiving. This claſs conſiſts of what are called the compound flowers,
and is certainly a natural one, if we except a
few genera· which are contained in the laſt
order, and which are placed in this claſs from
the ſingle circumſtance of having their anthers
united in a cylinder; one of theſe génera is
the viola, under which the violet and panſie
are ranked: this muſt be allowed to be a
fault in the ſyſtem; but at preſent it is our
buſineſs to conſider only the compound flowers:
Linneus makes the eſſence of a compound
flower to conſiſt in the union of it's anthers
into a cylindric form, one ſeed being placed
on the receptacle beneath each floret. A compound flower is ſo called from being compoſed
of many ſmall flowers or florets, which are
fixed on a common receptacle, and encloſed by
a common calyx. Theſe florets vary greatly
in their contents, the ſtamens and piſtils,
and alſo in the form of their corols, which
in ſome florets is tubular, in others flat, which
is called tongued. In the ſame flower ſometimes

times the border of the corol is wanting, and
fometimes there is not even a tube. On the
variety of form in the corol is founded, in
part, the generic character. On the florets
bearing ftamens or piftils, or both, are founded
the firft four orders. If all the florets of a
compound flower are found to contain ftamens and piftils, it muft then be referred to
the firft order: if fome of it's florets contain
ftamens and piftils, and others only piftils, you
muft look for your flower in the fecond order:
to the third it will belong if the florets in the
centre have both ftamens and piftils; and if
thofe in the circumference be deftitute of
either. The fourth order depends alfo on
the florets in the centre having both ftamens
and piftils; but from fome defect in the
piftils, producing no feed, the florets in the
circumference having only piftils, and producing feed. The fifth order is not diftinguifhed
by any circumftance belonging to the ftamens
and piftils, but is marked by the florets being
feparated from each other, and being enclofed in a partial calyx, all the florets being
contained in a common one, fo as to form
one flower. The character of the fixth order
is derived from the form of it's flowers being
fimple,

fimple, which perhaps ought to have excluded them from this clafs; but as they
agree with the compound flowers in the eſſential character of the united anthers, Linneus
has placed them in it; and as the principle
of the fyſtem on which he has founded his
claſſes does not pretend to make them natural,
there is not, perhaps, any great objection to
his having done fo; and while we receive fo
much amuſement from his arrangement of
the vegetable kingdom, we are bound to look
with candour upon any ſmall defects which
may appear in it. His life was ſpent in laborious reſearch into natural hiſtory, by which
the botanical world has been fo materially
benefited, that it ought at leaſt to pay the
tribute of gratitude to his memory. However, gratitude is not excluſively due to him;
much was done by his predeceſſors; and both
amuſement and inſtruction may be derived
from the ingenious fyſtem of Tournefort;
but at preſent we are to think only of Linneus as our great maſter. The characters of
the orders of the clafs Syngeneſia United Anthers, are too complex to retain in the mind
without having examined fome flowers belonging to them. The pupil fhould therefore

G   collect

collect a variety of the species arranged under thofe divifions, and, by diffecting them, imprefs upon his memory the different characters by which the orders are diftinguifhed. The dandelion (leóntodon), thiftle (cárduus), are proper flowers for inveftigation; it will be alfo expedient to examine fome violets and panfies as examples of the order of fimple flowers. There are fome flowers of the fourth clafs, Tetrandria, four ftamens, which are liable to perplex the young botanift in his fearch after compound flowers: in outer appearance the mode of infiorefcence in fcabious (fcabiófa) nearly refembles that of the compound flowers, although, on examination, there will be found very marked diftinctions between them. The fcabious, and feveral other génera of the fame habits, have their *four* ftamens feparate; the compound flowers, as is feen in the thiftle (cárduus), have their *five* anthers united in a cylinder: there is alfo another difference, thefe flowers of the fourth clafs have the florets, of which they are compofed, attached to the common receptacle by a fmall peduncle, or foot-ftalk; the florets of the compound flowers are feffile, or fixed to the common receptacle by their bafe, without

the

the intervention of a peduncle; the scabious, and that tribe of flowers which have not the essential mark of the United Anthers belonging to the compound flowers, are called aggregate. The flowers of both the thistle and dandelion, containing both stamens and pistils, refer them to the first order. Daisy (béllis), having the florets of the centre furnished with both stamens and pistils, and those of the circumference with pistils only, has a place in the second order. Blue-bottle (centauréa) has both stamens and pistils in it's central florets, and florets without either form the circumference; it is therefore of the third division. The fourth order not only derives it's character from the absence or presence of the stamens and pistils, but in addition to the necessity that the central florets should contain both, and the florets of the circumference only pistils, it is essential that the florets or the centre should be destitute of seeds, and that the florets of the circumference should be found to contain them; which circumstance distinguishes the fourth from the second order; and this distinction may be seen in the common marygold (caléndula) and daisy, which belong to those respective divisions.

The

The fifth order is readily underftod; each floret fhould be contained in a feparate calyx, and all together collected into one large common calyx; of this, globe thiftle (echinops) affords a fpecimen. The character of the fixth order confifts in the fingle circumftance of the united anthers, there being not one compound flower of this divifion. The ftigmas of the violet and panfie are worthy of obfervation: thefe flowers are both of the genus Viola, which is feparated into two divifions from the peculiarity of their ftigmas; that of common violet being reflected into a fimple hook, and that of the panfie (or three-coloured viola) being round and perforated. Jafione, or fheep fcabious, is placed in this order of fimple flowers, to which it certainly cannot belong, being compofed of many florets; nor is there any circumftance refpecting it's fructification, which gives it any pretence to be claffed with the compound flowers, except that of it's five anthers being flightly connected at their bafe, for they are not united in a cylinder: from the firft view of this plant it feems to be of the tribe called aggregate, but, on examination, it differs effentially from that order of plants in the numbers

of

of it's fructification and other circumstances.
The Jasione has proved perplexing, even to
proficients in botany; nor are the difficulties
which occur in it's construction yet explained
in a satisfactory manner.

There is a curious circumstance in regard to
the calyx of most of the compound flowers,
though not belonging to all, which is worthy
of attention. When the florets become mature, they burst open the common calyx, which
contains them; as soon as the stamens and
pistils of these florets have done their office,
they wither with the corols, the common calyx
then rises, and encloses the remaining parts
of fructification, till the seeds arrive at that
state of ripeness which makes them ready for
dispersion; the hairy down, by which they are
crowned, then expands, and again bursts open
the calyx, so as to bend it's leaves quite back,
and, by the help of this down, the seeds are
carried by the wind to a considerable diftance. Those compound flowers which have
their seeds furnished with a downy pappus,
take a variety of elegant forms; and the
class of United Anthers, though difficult at
first to study, amply repays our trouble in
attaining a perfect knowledge of it, from the

G 3   curious

curious mechanifm of it's flowers. The ftructure of the ftamens and piftils of the clafs
Gynándria, or twentieth clafs, is fo extraordinary as to be fuppofed by Linneus to occafion
the unufual appearance of the flowers belonging to it. The órchis tribe, paffion-flower
(paffiflóra), and árum, wake-robin, are of this
clafs; the effential character of which is the
ftamens growing on the ftyle, or on the receptacle elongated into the form of a ftyle,
bearing the piftil with the ftamens, and becoming a part of the piftil, which part muft
be well underftood before a diftinct idea of
the fituation of the ftamens can be obtained.
This clafs contains nine orders founded on
the number of ftamens in each flower. The
firft order, which is called Diándria, or two-
ftamens, is natural; the génera differing
from each other almoft only in the Nectary.
The ftructure of the fructification of this
order is very fingular; for the germe, always
beneath, is contorted: the petals are five,
of which the two inner converge, fo as to
refemble a helmet: the under lip conftitutes
the Nectary, which occupies the place of
the piftil and fixth petal: the ftyle grows to
the inner margin, and can fcarcely be diftinguifhed

tinguifhed with it's ftigma: the filaments are always two, very fhort, elaftic, and bearing two anthers, which may be divided like the pulp of a citron; they are enclofed in little cells opening downwards, and fixed to the inner edge of the Nectary; the fruit is a one-celled capfule, with three valves gaping at the angles. The genera of this firft order afford flowers which, in outward appearance, fo nearly refemble the animal kingdom, as to have occafioned a variety of fanciful names being given to them. The family of óphrys contains feveral fpecies, which refemble a variety of infects, the Nectary being the principal feature in their different forms; fometimes their flowers refemble a gnat, a butterfly, a bee, a fly, or a bird: the Nectary of the bee-óphrys is a large thick leaf of a footy colour, and, when feen in the light, feems varied with three bright yellow circular lines, with ruft-coloured fpaces between them, and fo exactly reprefents a drone, or bee, that it might be miftaken for them. The flowers of the genus Cyprepedium are fuppofed to refemble the form of a lady's flipper; and thence the plant has it's name. This curious tribe of flowers requires very accurate

G 4   inveftiga-

inveſtigation to enable us to underſtand their various parts, and affords much intereſting occupation to thoſe who take the pains to ſtudy it. The eight remaining orders of this claſs are known by their number of ſtamens. The ſtructure of the parts of fructification in the arum is moſt extraordinary, and not to be found in any other genus. The receptacle is enlarged into a naked club, with the germes at the baſe. The ſtamens are affixed to the receptacle, amidſt the germes, which is called by Linneus a natural prodigy: the moſt eminent botaniſts have been perplexed by this ſingular flower. The younger Linneus was of opinion, that every anther was to be conſidered as a diſtinct floret, and thence that the genus ought to be removed from the claſs Gynándria to the following one Monœcia, or ſtamens and piſtils ſeparate. I cannot pretend to decide on this ſubject, but hope, as this opinion of the younger Linneus opens a new principle of inveſtigation, ſome ingenious botaniſt of the preſent age may be able to diſcover the ſecret of the wonderful mode of fructification found in this family. An engliſh botaniſt ought certainly not to remain ignorant of a plant which contributes ſo much to the beauty

of

of our hedge-banks during the period of flowering, and continues to attract his eye by the brilliancy of its scarlet berries through most of the months of autumn. The following class, Monœcia, the twenty-first class, contains such plants as have their stamens and pistils in separate covers, but growing on the same root; hazle (córylus), nettle (urtíca), are instances of the Monœcia class, or class of one-house : the orders of this class are derived from the number, union, and situation, of the stamens, circumstances which constitute the chief characters in the classes, where the stamens and pistils grow together in the same cover. There are eleven orders of the class one-house, which are distinguished by the same names that are given to the preceding classes. Hazle (córylus) having several stamens in each scale of it's ament, or catkin, is placed in the order Polyándria, many stamens; nettle (urtíca) in Tetrándria, four stamens; and cypress (cupréffus), which is also of this class, is arranged under the order Monadélphia, one-brotherhood, having it's stamens united at their bafe, like the flowers of that class, which might lead a young botanist to place it among them if he did not keep in

his

his mind the effential circumftance of the firft twenty claffes, viz. their having their ftamens and piftils in one flower. To this clafs of one-houfe belongs the nutmeg (myríftica), for the knowledge of which flower the world is indebted to Dr. Thunberg, who has given a defcription of the genus from the real flowers, whereas the former characters were taken from a plant which had no affinity to the true nutmeg. The clafs Diœcia, or two-houfes, contains thofe flowers which have their ftamens growing on one plant, and their piftils on another. Vallifnéria belongs to this clafs: the wonderful progrefs of the flowers of this plant feems to furnifh a ftrong argument for the fenfation of plants; but this is not the time to enter into the difcuffion of that part of our fubject. Hemp (cánnabis), hop (húmulus), mercury (mercuriális), and willow (fálix), all belong to the clafs two-houfes: there are fifteen orders contained in this clafs, characterized from the fame circumftances with thofe of Monœcia, or one-houfe, and named by words expreffive of thofe circumftances. Great fault is found with the contradictions that this occafions; and certainly this part of the fyftem is open to cenfure,

3

and

and in all probability would have been corrected, had Linneus's health, during the latter
part of his life, permitted. Alterations have
been made in thefe claffes of late years, which
are pretty generally received; and as the liberal
fpirit of the age inclines his fucceffors in this
delightful fcience rather to render his labours
perfect, than to hold out his failings to ridi-
cule, we may hope that time will give us his
fyftem as free from defect as fuch an undertaking can be expected to be.

The mifletoe (vifcum) belongs to the clafs
two-houfes: this is a parafitical plant, or one
which lives upon the juices of another vegetable, without fixing it's roots into the
ground: it can only be propagated by ftick-
ing the feeds upon the bark of trees, into
which they ftrike their roots in a curious
manner. A feed firft fends out three claws,
which fix themfelves on the bark of the tree,
and begin to feparate at the centre of the
feed, as if each claw was to become a diftinct
plant; but in a year or two the three claws
become fwoln and enlarged enough to meet
at their points, and are fo ftrongly united,
that they make the foundation but of one
plant; the place of their firft joining in the

centre

centre opens and divides, fo that three diftinct branches appear fpreading from the root; after this, it proceeds to bloffom and bear fruit, and will live to a great age, agreeing very well with it's fofter tree, which it ornaments, in grateful return for the fupport which it re-ceives: it grows moftly on apple-trees, but is fometimes found on the oak, though rarely, and on feveral other kinds of trees; the feeds are enclofed by fo vifcous a pulp, that they readily adhere to other vegetables, on which they are frequently dropped by birds, and thus the fpecies is propagated.

EXPLANATION OF PLATE IV. PART I.

OF THE CLASSES.

Fig. 1. One Stamen, Monándria.

Fig. 2. Two Stamens, Diándria.

Fig. 3. Three Stamens, Triándria.

Fig. 4. Four Stamens, Tetrándria.

Fig. 5. Five Stamens, Pentándria.

Fig. 6. Six Stamens, Hexándria.

Fig. 7. Seven Stamens, Heptándria.

Fig. 8. Eight Stamens, Octándria.

Fig. 9. Nine Stamens, Enneándria.

Fig. 10. Ten Stamens, Decándria.

Fig. 11. Eleven to Nineteen Stamens, Dodecándria.

Fig. 12. Not lefs than Twenty Stamens placed on the Calyx, Icofándria.

Fig. 13. Many Stamens placed on the Receptacle, Polyándria.

Fig. 14. Two-powers, Didynámia.

Fig. 15. Four-powers, Tetradynámia.

Fig. 16. One-brotherhood, Monadélphia.

Fig. 17. Two-brotherhoods, Diadélphia.

Fig. 18. Many Brotherhoods, Polyadélphia.

Fig. 19. United Anthers, Syngénefia.

Fig. 20. Stamens on the Piftil, Gynándria.

Fig. 21 One houfe, Monœcia.

Fig. 22. Two-houfes, Diœcia.

Fig. 23. Polygamies, Polygámia.

Fig. 24. Fructifications concealed, Cryptogámia. *a.* Fern, *b.* Mofs, *c.* Lichens, *c** fringed Lichen of the natural fize, *c.* the fame magnified, *d.* a fungus.

Plate 4. *Part I. P. 94.*

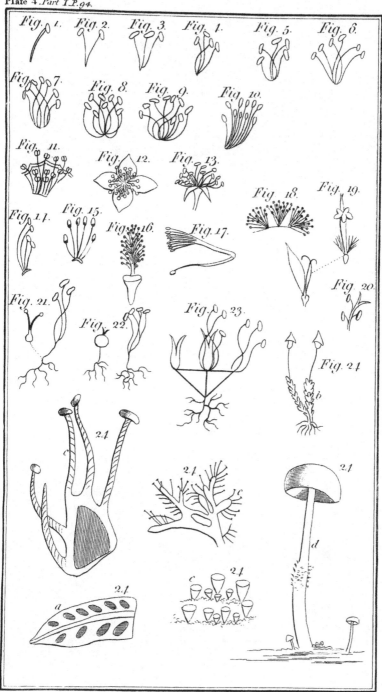

LECTURE V.

Clafs Polygámia explained; Caprification. Clafs Crypto-
gámia explained,

THE effential charaĉter of the clafs Polygámia confifts in the plants, of which it is
comprifed, having, on the fame root, flowers
which contain ftamens and piftils within the
fame cover, and alfo other flowers, which
bear either ftamens feparately, or piftils feparately; fometimes flowers are found on the
fame plant, which contain ftamens and piftils,
ftamens without piftils, and piftils without
ftamens: the prefence of the firft kind marks
the clafs; without flowers, which contained
both ftamens and piftils, the plant would belong to either the clafs one-houfe, or two-
houfes. The plants of the Polygamia clafs
are many of them difperfed, by botanic
writers of the prefent age, into Monœcia
and Diœcia; fo that probably that clafs will
foon be banifhed from the fyftem. The
orders, of which there are three, depend on
the

the difpofition of the ftamens and piftils in the flowers of the different plants. The fig (ficus cárica) long perplexed the botanic world, to difcover by what mode the duft of the ftamens could be conveyed to the piftil, as thefe parts of fructification are enclofed within feparate fruit, this fruit not being a feed-veffel, but a receptacle furrounding the ftamens and piftils, which grow upon it; and fome of them fo clofely immured, that the manner in which they are fertilized was incomprehenfible. At length it was difcovered, that a kind of gnat depofited it's eggs in thefe receptacles, and, by going from one kind of fig to the other, was fuppofed to bear on it's wings the anther duft of the ftamen-bearing fig to the ftigmas of that which contained only piftils. This procefs performed by the gnat was called caprification, and was fo ftrongly believed to be effential to the ripening of the cultivated fig, that the inhabitants of the Archipelago, who trade with their figs, fpent much time in obferving the critical moment of the gnat iffuing out of one kind of fig and entering the other, and fometimes gathered the fruit, in which the gnat was contained, and brought it to that which they wifhed to

have

have fertilized. Mr. Milne gives a long and
curious account of the procefs of caprification;
but it is difficult to affent to the truth of the
neceffity of it, there appear to me fo many
objections againft it. Firft, there is not any
fpecies of fig known, which bears piftils only;
confequently not any which is not fufficient
in itfelf to it's own fertilization. In Provence
and Spain the cultivated fig is proved to be
fo by being brought to perfection without the
procefs of caprification. Secondly, thefe fruits
generally open at the top at the time that
their ftamens become mature; a circumftance
analogous to all water plants, which rife to
the furface, when their ftamens are ready to
fcatter their duft, in order that they may difperfe it in the open air; an element which
feems neceffary for that procefs.

The procefs of caprification has been
efteemed a powerful argument for Linneus's
fyftem of the anther-duft being effential to
the perfect production of feed, and made ufe
of as fuch by many intelligent authors. The
late ingenious Dr. Darwin found fo many
difficulties to be furmounted in the belief of
this procefs, that he ventures to refufe his
affent to it. He conjectures that thofe figs,

H   which

which have their receptacles clofed on all
fides, might be vegetable monfters cultivated
for their fruit, as thofe grapes and barberries
are, which are without feed; and that the
procefs of caprification might be of imaginary
ufe, or that it might contribute to ripen the
fruit, as thofe apples ripen fooner which are
wounded and penetrated by worms in our
own climate; and this feems probable from
what is told us by Mr. Milne concerning the
figs of Malta; one kind of which, he relates
from Tournefort, bears two crops in the fame
year, the figs of the firft being fweet, and
arriving at perfect maturity *without* the affiftance of caprification; thofe of the fecond
being much fmaller, and not ripening at all,
if this procefs be not followed. Tournefort
adds, that the figs in Provence and in Paris
ripen fooner if they are pricked with a ftraw
dipped in oil, which feems to make it probable that the puncture of infects in caprification may caufe the fecond crop of fruit to
arrive earlier at maturity in Malta; that is,
before the inclement part of the feafon comes
on; as in our climate the plums and pears
wounded by infects frequently ripen fome
weeks fooner than the others, to which that
circumftance

circumftance has not occurred. The fig-trees cultivated in our own country produce two crops; the firft upon fhoots of a year's growth, which appears in fpring, and arrives at maturity in the courfe of the fummer; the laft crop does not put forth till autumn, and proceeds from the fhoots of the preceding fummer. This crop can never ripen in our climate, and is carefully pulled off by the gardeners. It would feem that the tree has not power to bring two crops to perfection, even under the influence of more benignant fkies, as at Malta, as the fruit obtained by the procefs of caprification is fcanty and of bad quality.

The neceffity of this operation has, however, univerfally obtained belief in the eaft; but, in this inquiring age, we cannot eafily affent to facts to which we think both reafon and analogy oppofed. If a fig be cut open at the time when it gapes at the top, the florets may be feen arranged on the infide in a beautiful manner, and there may be found feveral of the ftamen-bearing kind in the ftate of difperfing their duft.

We are now arrived at the twenty-fourth or laft clafs of the Linnean fyftem, the clafs

Crypto-

Cryptogamia, or clandeftine marriage, the grand defideratum of botany, as the plants of which it confifts have their fructification fo obfcure, that there are but few génera in which it has yet been diftinctly feen. This clafs includes all thofe plants, which have a ftructure different from thofe comprifed in the other three and twenty claffes, and is divided by Linneus into four orders, the filices, ferns; mufci, moffes; algæ, wrack, or feed-weed; fungi, fungufes. The little knowledge, that has hitherto been obtained of thefe numerous tribes of plants, has been confidered a great reproach to the fcience of botany. Perhaps the fyftem of Linneus may have retarded a more diftinct arrangement of them, that being founded upon the parts of fructification, which in moft of the génera belonging to the clafs Cryptogamia are fo difficult to afcertain. The ferns are defined to be plants bearing their flowers and fruit on the back of the leaf or ftalk, which in this tribe of plants are the fame, the ftem not being diftinguifhable from the common foot-ftalk, or rather mid-rib of the leaf: fo that, in ftrict propriety, the ferns may be faid to be without ftems. The ftem and leaf thus

united

Plate 5. *Part I. P. 101*

Vegetable Lamb

Polypodium Barometz.

London Published May 1. 1797. by J. Johnson S. Pauls Church Yard.

united are termed by Linneus a frond. The feed of the ferns affords an inftance of the moft curious mechanifm, and will be well worthy the attention of proficients in botany. All that is neceffary for the pupil in that fcience is an acquaintance with an outline of the characters of the génera contained in the clafs Cryptogamia, of many of which a clear idea may be obtained by ftudying plates of their extraordinary ftructure given by various ingenious artifts. The true fago powder is faid to be made from the pith of a fpecies of fern, *Cy'cas circinalis*; and that great vegetable curiofity, the tartarian lamb, is now known to be the root of the polypodium barometz, which, being pufhed out of the ground in it's horizontal fituation by fome of the inferior branches of the root, bears fome refemblance to a lamb ftanding on four legs, which is increafed by the thick yellow down, by which it's root is covered. And, indeed, ftories fo extraordinary of the appearance of this fern have gained admiffion into the works of authors of fo much repute, as to have given the tale a degree of credibility far beyond it's deferts.

Many things have gained the character of monfters from want of that inveftigation,

which

which ought always to be given to hiftories of a marvellous kind. In former ages we might probably have received from travellers a grave account of a tree, bearing gloves, and ftockings, and caps, growing in Caffraria; the report of which was fo general as to excite the attention of Dr. Thunberg, when travelling in that country. With his ufual affiduity he unveiled this myftery, and found all this wearing apparel to be nothing more than the downy leaves of the Bupléurum giganteum, which, by a little dexterous management, were converted into thofe various articles, which were afferted to grow upon the plant.

In fome countries the roots of different fpecies of fern are ufed in the procefs of making bread. Captain Cook relates, that in New Zealand the common fern (pteris aquilina is chofen for that purpofe. Bread is alfo made from a fpecies of fern by the inhabitants of Palma, one of the Canary ifles, when corn is fcarce, and is faid to be little inferior to that made from wheat.

But to proceed to the fecond order of Cryptogamia. The moffes (mufci) are divided according to their anthers, being calyptred, or not calyptred, being on the fame, or feparate

plants,

plants, and having the piftil florets folitary, or growing in cones. Their feeds have no co-tyledons, or any proper coverings. Linneus doubts, whether what he has called anthers might not, with greater propriety, take the name of capfules, and their duft be confidered as true feeds, as in Buxbaúmia, and fome other génera, have been feen within the covers real duft-bearing anthers depending from their filaments, gaping at the top to difcharge their duft on the fringes, as on piftils. Dillenius, profeffor of botany at Oxford, was the firft who attempted an arrangement of the moffes. There are many curious circumftances be-longing to the tribe of moffes, one of which is their having this fingular property, that, though preferved dry for feveral years, upon being moiftened they refume their original verdure, and probably their power of vege-tation; an experiment eafy to be made. The fructification of the flags, or algæ, is fo obfcure as not to admit of precife arrange-ment; they are only divided into terreftrial and aquatic, and the génera diftinguifhed by their outer ftructure. This order contains many curious and ufeful vegetables; among the latter there is none more worthy of notice than the lichen rangiferinus. This little plant

may

may be properly efteemed the fupport of millions of mankind, as it is the fole food of the rein-deer; without which ferviceable animal, the inhabitants of the northern regions could not exift. The rein-deer furnifhes them with milk, butter, and cheefe, draws them in fledges with eafe and fwiftnefs over vaft tracts of land buried in fnow; his flefh affords them food; his fkin, clothing; his tendons, bow-ftrings; and his bones, fpoons. All thefe benefits would be loft, had not nature formed this lichen fo as to enable it to vegetate beneath the fnow, by which it is commonly covered to a great depth: the rein-deer, however, contrive to dig through the fnow with their feet and brow-antlers, till they arrive at their food. To the common name of rein-deer lichen, by which this plant is known, it has therefore the fulleft claim. The whole tribe of lichens poffefs qualities of which various ufes are made; different fpecies being ufed in dying reds and purples. Dr. Thunberg relates, that the Japanefe gather a fpecies of ulva, which is one of the algæ, and, clearing it from all impurities, dry and reduce it to a fine powder, which they eat with boiled rice, and fometimes put into foup. There are other fpecies alfo of them, which are ufed for

food

food or pickles by ourſelves. The formation
of ſome of the génera, which belong to the
aquatic diviſion of this order, is worthy of
remark. The conferva ægagrópila is of a
globular form, from the ſize of a walnut to
that of a melon, much reſembling the balls of
hair found in the ſtomachs of cows. It does
not adhere to any thing, but rolls from one
part of the lake, on which it lives, to another.
The confÉrva vagabunda has it's name from it's
wandering habits. It dwells on the european
ſeas, travelling along in the midſt of the waves.
Theſe may not improperly be called itinerant
vegetables. In the ſame manner, the fucus
natans ſtrikes no roots into the earth, but
floats on the ſea in extenſive maſſes, and may
be ſaid to be a plant of paſſage, as it is wafted
by the winds from one ſhore to another. The
byſſus flos-aquæ, water flower, oats on the
ſea all day, and ſinks a little during the night,
as if to protect itſelf from the injuries of noc-
turnal air ; or poſſibly this may be it's mode
of ſleeping or taking reſt.

The changes of appearance in confÉrva polymórpha are moſt extraordinary, and have
given riſe to ſome beautiful lines in the Botanic
garden. This plant twice changes it's colour
from red to brown, and then to black, and va-

ries

ries it's form, by lofing it's lower leaves, and lengthening fome of it's upper ones, fo as to be miftaken by unfkilful botanifts for different plants: it grows on the fhores of this country. The laft order of the clafs Cryptogamia confifts of the Fungufes, or Fungi. Linneus has divided this order of plants according to the method of Dillenius; indeed he does not feem himfelf to have attended to any of the orders of this obfcure clafs, with that indefatigable refearch, which characterizes his labours in regard to the other part of the vegetable kingdom; but, with a candour belonging to true knowledge, he frankly owns himfelf indebted to Dillenius, and Micheli, for the information he is able to give the world refpecting them. The method of Dillenius, which Linneus has followed, is founded upon the figure of the Stipe, or Foot-ftalk; the hat, or upper part, with it's plates, holes, and cavities; and from the variety of ftructure in thefe parts, has divided the whole Fungus tribe into ten Génera. The fudden appearance of thefe kinds of plants, in places where they had not been known before, gave rife to the belief, that they had their origin from putrefaction; but this has been clearly proved to be a miftake, and that they are produced

from

from feeds; that their fpecies are conftant, and renewed by uniform laws; notwithftanding it muft be confeffed, that we are yet much in the dark concerning this part of the vegetable creation; but, as it is now particularly attended to, a few years may probably make us acquainted with the various modes of it's reproduction. We already owe much to the accurate inveftigations of Mr. Curtis, and to other able botanifts of the prefent age, who have elucidated the knowledge of thefe plants by many beautiful drawings. In the clafs Cryptogamia advantage may be particularly derived from thefe publications, as by ftudying the pictures of various plants belonging to that clafs, an intereft in the originals will be acquired, and the ftudent be led to fearch into their hiftories, in which, no doubt, there is much curious matter to be acquainted with. The late difcoveries of the wonderful manner by which various fpecies of the animal kingdom are continued, may poffibly lead to fome equally extraordinary in the modes of vegetable reproduction. The hiftories of the Polypi or Hydræ aftonifh us, particularly of the Hydra Stentorea, which multiplies by fplitting lengthways; in twenty-four hours the two divifions, which adhere to a common pedicle. re-

fplit,

split, and form four distinct animals; these four
in an equal time again split also, and thus proceed, doubling their numbers daily, till they
acquire a figure somewhat resembling a nosegay; the young afterwards separate from the
parent stock, attach themselves to the roots or
leaves of aquatic plants, and each individual
gives rise to a new colony. The fresh-water
polypus may be cut into innumerable divisions,
and every separate piece will become a separate animal; a history so analogous to the
tale of the hydra's heads, as to induce us no
longer to believe that story fabulous; and indeed we have facts from the experiments of
Monf. Trembly in regard to the fresh-water
polypus, or hydra, which equal any ideas
that could occur to the most romantic fa-
bulist. And may it not be found, in some of
the tribes of vegetables belonging to the
class Cryptogamia, that similar modes of increase take place, exclusive of all others? for
the increase of plants by strings and suckers,
may be considered analogous to the reproduction of the Hydra genus. On so obscure
a subject light might, perhaps, be thrown
from experiments founded on analogy: it is
certain that little progress has been made in
the knowledge of these extraordinary vegetables

by

by thofe who have proceeded upon the ex-
pectation of the ufual mode of fructification.
The uncommon beauty of an affemblage of
thefe plants on our banks, walls, and heaths,
in winter, muft engage the attention of every
botanift. There is a fpecies of fungus, the lycoperdon fornicatum, or turret puff-ball, which
is of a very extraordinary form, having the
appearance of an inverted mufhroom. The
plate here given of this fingular vegetable was
taken from a peculiarly fine fpecimen found
growing in the kitchen-garden of Mr. Rook,
near Mansfield.

Adjoined to the claffes is an appendix confifting of plants, which Linneus rather chofe
to place apart than to diftribute into the feveral claffes of his fyftem, and this on account
of their fingular ftructure: he has arranged
them all under the head of Palms, and defines
them to be plants with fimple ftems, bearing
at their fummit leaves refembling thofe of
ferns, which are termed Fronds, and are a
compofition of a leaf and a branch. Their
flowers and fruit are produced on that particular kind of receptacle called a fpadix, protruded from a common calyx in form of a
fheath, termed by Linneus a fpathe. The
terms

terms fpathe and fpadix were originally applied to palms only, but are now ufed with
much greater latitude, and applied to the nar-
ciffus, árum, and many other plants, the flowers
of which are protruded from a fheath. In the
palms the fpadix is branched, in all other
plants it is fimple, admitting of fome variety
in the difpofition of the flowers. The cocoa-
nut-tree (cócos nucifera) is a palm, fo is the
date-tree (phœnix dactylifera); and it is afferted
by fome authors, that if the ftamen-bearing
flowers of this plant are gathered in a proper
ftate of maturity, and dried, the duft of the
anthers will retain it's virtues for more than a
year; the fame alfo is faid of the piftacia,
which belongs to the clafs two-houfes (Dioecia); the cory'pha umbraculifera belongs to
this majeftic order of vegetables, being often
200 feet in height: it is a native of the Weft
Indies, and has obtained the name of umbrella-
bearing, from the fhelter which it's large
feathered leaves afford to the inhabitants of
that fcorching climate from the ardent rays of
the fun. This tree has alfo been called the
cabbage-tree, but erroneoufly: Mr. Forfter informs us, that the true cabbage palm is a
fpecies of aréca, the aréca oleracea, fo called,
probably,

probably, from the use that is made of the kernel-like substance, which is found towards the top, and which is a most grateful and salutary food to sailors, who have been long confined to salt diet; on which account, this substance has been celebrated by all navigators, and from them has obtained the name of cabbage, from it's resemblance in taste to that vegetable. Some writers have mentioned it as being commonly made use of for food by the inhabitants of the countries where this palm-tree is found: but this must probably be an errour, as, from the best authorities, it appears that the kernel-like substance, or cabbage, is esteemed a rarity even in the West Indies, and frequently pickled and sent to England as a peculiar nicety, although the tree is a native of the soil. Nor is it difficult to account for this scarcity when we attend to the fact, that the part called cabbage cannot be obtained but by the destruction of the whole tree; nor will this appear extraordinary if we consider the mode of it's structure: the whole tribe of Palms bear their leaves on the upper part of their stems only, some of which rise to the height of 200 feet; the part eaten as cabbage seems to be the yearly shoot, by

I   cutting

cutting off which the leaves, which fhould
form the buds for the enfuing year, are deftroyed, and with them the life of the plant.
If the leaves from any common tree are
ftripped off, fo as to prevent the formation of
buds, the tree will be cither killed, or it's vigour fo far deftroyed as to render it of no
value.

Although the aréca oleracea is the only
palm which bears the cabbage part in great
perfection, the cocoa-nut palm, and feveral
other kinds of palm, are faid alfo to afford
it; but the accounts of this tribe of vegetables
are often fo fhort, and given in a manner fo
confufed, that there is hitherto little accurate
knowledge obtained of their habits. The
hiftory of the vegetation of the tropics, by a
philofophical botanift, would be a work of
the firft value. There is another tree, which
is known by the name of the Bread-fruit
tree, which is an inhabitant of the iflands of
the South-Sea, and alfo of afiatic growth; of
much more extenfive utility than the cabbage-
palm. This is the artocárpus commúnis of
Forfter, and belongs to the clafs Monœcia,
one houfe. The various attempts which have
been made to introduce this valuable tree into
the

the West India iflands promife at length to be
fuccefsful. There are now plantations of it in
Jamaica, from which fruit has been gathered.
Nearly twenty years ago Dr.Thunberg exerted
his beft endeavours to bring it into Europe;
but at the time, when he flattered himfelf that
he was on the eve of depofiting his treafure
with fafety, all his hopes were fruftrated by a
violent ftorm, which endangered the lofs of
the veffel on board which he was conveying
his valuable cargo of more than a hundred
bread-fruit trees, and other rare plants, all of
which were deftroyed. Thefe trees he had
brought from the ifland of Ceylon, the inhabitants of which make ufe of the fruit in a
variety of luxurious difhes. Dr.Thunberg enumerates fifteen different ways in which they
have it prepared; but that which gives this celebrated tree it's real importance is the extenfive
benefit which is derived from it to the poorer
claffes of the people, who make ufe of it's
fruit to fupply the place of bread or rice, or
as our poor do of potatoes, whence it's name
of bread-fruit. The natives of Otaheite, of
all degrees, make ufe of it alfo in the moft
fimple manner; they bake it amongft hot
ftones for food, and mix it with water for

I   their

their liquor. There are two kinds found in Ceylon; one which yields smaller fruit, has no seeds, and is more rare; the other, bearing fruit from thirty to forty pounds weight, grows in all parts of the island, and produces feeds to the number of two or three hundred, each of which is four times the size of an almond. Mr. Forster tells us, that the bread-fruit tree of the South-Sea isles has four or five varieties, all without feed; which deficiency he attributes to the effects of cultivation; but as Dr. Thunberg, contrary to his usual accuracy, omits giving the botanical names of the bread-fruit tree of Ceylon, it cannot be ascertained in what particulars it differs from, or agrees with, those of the Pacific Ocean; but there can be little doubt that they are of the same genus. If they are deprived of their feeds by cultivation, they lose a part which in Ceylon is much esteemed as a nutritious and palatable diet, these feeds being prepared for the tables of the rich in different ways. Fried in cocoa-nut oil they are esteemed a great delicacy; by the poor they are eaten roasted like chesnuts, alone, or mixed with the pulpy part of the fruit, which they also frequently eat simply boiled

or

or roafted, or fometimes mixed with a little
rice, rafpings of cocoa-nut, onion, and a
fmall quantity of falt and turmeric. The
bread-fruit trees flourifh for whole centuries,
and bear their fruit, which ripens by degrees,
not only upon the thickeft branches, but
upon the ftem itfelf, for the fpace of eight
months together. The fruit is ufed for food
in three different ftates of ripenefs, but cannot
be eaten without preparation, till it arrives at
maturity; at which time the pulp, which
furrounds the feeds, has a fweetifh tafte, and
is often eaten in it's frefh ftate, after peeling
off the rind, which is thick, and covered with
prickles.

The banana and plantain tree (mufa fapien-
tum, and paradifiaca) natives of the Weft-
Indies, have obtained the name of bread-trees
from the fame caufe that the artocárpus has
been fo called; many hundred acres of them
being cultivated in Jamaica for the ufe of the
negroes, who are faid to prefer the fruit of the
plantain tree, when roafted, to bread, and that
moft of the native whites ufe it in the fame
manner. The banana is alfo found in the South-
Sea ifles, and is faid by Mr. Forfter to lofe it's
feeds by cultivation, as the artocárpus does;

I 2   but

but it is not food only that thefe trees fupply
to the inhabitants of the warm climates: the
banana adminifters to their wants by the
fhade of it's leaves, the fize of which is often
eight feet long, and three feet broad. It is
moft interefting to read the accounts given of
the vegetables in thofe luxuriant regions,
which thefe trees, among others of equal or
more extenfive ufe, inhabit. The cocoa-nut
tree feems to merit a place in the firft rank;
and Dr. Thunberg tells us of two fpecies of
palm-tree in Ceylon, the boraffus flabelliformis, and licuála fpinofa, the leaves of which
are ufed without any further preparation than
feparating and cutting them even, for writing
upon; the method of performing which is to
carve with a fine pointed ftyle the letters upon
the leaf, and then rub them over with a fine
charcoal, which gives them the appearance of
having been engraved: thus they write all
public edicts and letters, and form books by
ftringing feveral flips of thefe leaves together,
and ornament them by figures engraved in the
fame manner as the letters: one of thefe books
Dr. Thunberg brought with him to Europe.
The leaves of the licuála palm are alfo ufed for
umbrellas; one fingle leaf is faid to be fufficient

3   to

to fhelter fix perfons from the fun or rain; a luxuriancy of vegetation of which europeans can form but very inadequate ideas.

Linneus has annexed to his Génera Plantárum an attempt to arrange all known vegetables according to their natural affinities; which, from the principle of his artificial method, are neceffarily feparated, and diftributed amongft the various claffes in his fyftem. To eftablifh a natural method, or one founded on the numerous, permanent, and fenfible relations, that one plant bears to another, has been attempted by many eminent botanifts, and with much fuccefs in regard to many of the génera; but, unlefs the fpecies could alfo be arranged in the fame manner, a fyftem cannot be eftablifhed upon thefe principles. The fuperior excellence of an artificial fyftem feems now to be generally allowed, as more readily leading us to the knowledge of a plant, that we may wifh to be acquainted with, fo far as it's clafs and order. However, Linneus was of opinion, that time would difcover a natural fyftem; and that all plants, of what order fo ever, would be found to fhow an affinity to fome others, to which they are nearly allied; and on this principle

I 3   he

he has arranged his natural orders, of which there are fifty-eight, and rather more than a hundred génera, which he calls yet dubious. Thefe orders are well explained in Mr. Milne's Botanical Dictionary, where we will ftudy the characteriftic marks by which the plants contained in them are affembled; but a complete knowledge fhould firft be obtained of the artificial fyftem, which will enable the pupil to diftinguifh plants, and he may then proceed to the natural orders, where he may learn the nature of them.

BOTANICAL LECTURES.

PART THE SECOND.

LECTURE I

Génera of Plants.

HAVING acquired the knowledge of the seven parts of Fructification, of the various modes of Inflorescence, and of the Classes with their Orders, the pupil may begin with the Génera of plants, or third division of the system. A Genus is an assemblage of several species of plants, which resemble each other in their most essential parts, and has often been well compared to a family, the whole of which bears one common name, while a particular one, or a specific name, is given to each individual. Linneus has demonstrated, that nature has imprinted certain characteristic marks on

I 4

the

the parts of fructification, which may be esteemed the alphabet of botany, and by the study of which alphabet we may learn to read the génera. He enumerates 26 marks or letters; the first six are taken from the calyx. 1st, the Involucre; 2d, the Spathe; 3d, the Perianth; 4th, the Ament; 5th, the Glume; 6th, the Calyptre; three from the corol, the Tube and Claws, forming the 7th character; the Border the 8th; and the Nectary the 9th. The stamens afford two marks, 10th, the Filaments, 11th, the Anthers. The pistil three; 12th, the Germe; 13th, the Style; 14th, the Stigma. From the Pericarp are derived seven; 15th, the Capsule; 16th, the Silique; 17th, the Legume; 18th, the Nut; 19th, the Drupe; 20th, the Berry; 21st, the Pome. From the seed are taken two; the Seed itself the 22d mark; and the Crown the 23d. The Receptacle of the Fructification makes the 24th; the Receptacle of the Flower the 25th; and that of the fruit the 26th, which completes the alphabet. These two kinds of receptacles may require some explanation. The receptacle is that of *the fructification,* when it contains the corol, the stamens. the pistils, and the germe, which belong to one flower.

When

When it is a bafe, to which the parts of the
flower are joined, and not the germe, it is a
Receptacle *of the flower*, which may be feen
in dog-tooth violet (dens canis), primrofe
(primula), and in various other flowers: in
which cafe the germe, being placed below
the receptacle of the flower, has a proper
bafe of it's own, which is called the Receptacle of the Fruit: of this the tree-primrofe
(cenóthera) is an example. Linneus does not
mention the Receptacle in his Génera Plantarum, except when he can introduce it as a
character varying in fhape and furface; by
which feveral of the génera of the clafs Syngenéfia, United Anthers, are diftinctly marked.
With the alphabet, or 26 marks taken from
the fructification, added to the number,
figure, fituation, and proportion, Linneus
has fo well diftinguifhed the génera from
each other, that nothing more is wanting to
enable us to read the whole vegetable kingdom. When an effential character could be
obtained he has added it, as that taken from
the nectaries in parnáffia, héllebore, ranúnculus, and áconite. Could fo diftinguifhed a
mark be found in all génera, it would render
the ftudy of botany agreeable indeed; and we

are

are not to defpair of time bringing about this much wifhed for improvement; and it more probably will be obtained, if we content ourfelves with making the principal point of our labours the perfecting the fyftem of our great mafter, than if we endeavour after fame by feeking to eftablifh a new one. In the firft attempts of the botanical pupil to refer his flowers to their proper génera, fome difficulties may occur, and he may find the language of the tranflated fyftem of vegetables uncouth to his ear; a very fhort time, however, will render it familiar, and he will then perceive the fuperior excellence of it's expreffive concifenefs over every other work which has yet been publifhed for the ufe of the englifh botanift. The canna indica, a plant to be found in all hot-houfes, and the hippúris, mare's tail, with which our ditches abound, are proper fpecimens for examination. Thefe flowers, containing each one ftamen and one piftil, muft be looked for in the firft clafs and order Monándria Monogynia. On opening the book at this clafs, the pupil will find the names of thirteen different plants; thefe plants are feparated into two divifions; in the firft divifion there are ten plants, the
character

charaƈter of which is " fruit celled, *beneath*."
The terms *beneath* or *above*, applied to the
germe, expreſſes it's ſituation in regard to the
receptacle. In the roſe it is below, alſo in
apples; and the ſame ſituation of the ſeed-
veſſel being made uſe of as a mark by which
the ſubdiviſion of an order is diſtinguiſhed,
the neceſſity is evident of becoming acquainted with theſe very minute peeuliarities.
Under the ſecond diviſion, charaƈterized by
" fruit celled, one-ſeeded," there are three génera; at the ſame time the names of two
other plants occur, printed in italics, valeriána
rubra, and calcitrápa, which may require ſome
explanation: theſe are two ſpecies of valeriána, which have but one ſtamen. When
Linneus has thought proper to make the circumſtance of an individual plant differing in
the number of ſtamens from the reſt of it's
genus, the mark of the ſpecies, he has always
noted ſuch plants under the claſſes to which,
in ſtriƈt propriety, according to the rule of his
ſyſtem, they ſhould have been referred, and
marked them with an aſteriſk; ſo the lychnis
dioica will be found noted in the claſs two-
houſes; and ſeveral others in the ſame
manner.

The

The character " fruit celled, beneath," places the canna in the firſt diviſion of plants of the firſt order From the firſt ſix it differs ſo materially in appearance that there can be no doubt in rejecting them; but to the ſeventh, kœmpféria, there is ſome ſimilarity; the corols of both are " ſix-parted, lips two-parted." The revolute form of the corol diſtinctly marks the canna. The genus being diſcovered, the number by which it is marked muſt be obſerved. Canna is diſtinguiſhed by No. 1; by turning over the page that number will be found, and under it a more diffuſe deſcription of the character of the genus. The hippúris there can be no difficulty in diſcovering; it's ſingle ſeed ranks it plainly under the ſecond ſubdiviſion of the firſt order, to which it's one piſtil had referred it: it will be found deſtitute of calyx and corol, marks which diſtinguiſh it from the two other génera with which it is arranged. The No. 11 refers it to a fuller deſcription, which ſo well agrees with it's habits, that it's genus cannot be doubted of. Thus through all the claſſes the ſame method of arrangement will be found; a method which greatly facilitates the ſtudy of the plants contained in them, and

particularly

particularly of thofe claffes wherein very many génera are comprifed. The different fpecies are alfo arranged in the fame manner, when any peculiar character occurs in a certain number of them, as in lonicéra. When the young ftudent has gathered a honeyfuckle, he muft firft examine it's claffical character: he will find five ftamens, with one piftil; which parts of fructification will refer the plant to the clafs and order Pentándria Monogynia. He muft then examine the fubdivifions of that order, and will find that his flower muft belong to that which is characterized by " flower one-petalled, *above*;" the term *above* expreffing that the germe is beneath the other parts of fructification. Under this divifion he will meet with between thirty and forty génera; but perceiving that the feed-veffel is a berry, he will find his fearch limited to not more than twelve. The number of feeds within the berry, or the number of cells which it contains, are not obvious characters to an unexperienced eye; the form of the corol, however, is evident to the moft fuperficial obferver; and there are only two génera in which they are marked as unequal; thefe are the lonicéra and the triófteum, and

between

between thefe two there is fo clear a diftinction in the form of their ftigmas as cannot
be miftaken, that of the lonicéra being
headed, and that of the triófteum oblong.
The more diffufe account of the genus muft
ftill be inveftigated. The number of lonicéra
is 233, which refers to the fame in the fuller
defcription of the genus: this defcription
agrees with the charaćter of the honeyfuckle.
Again: under the generic charaćters there are
three divifions; thefe divifions are of the fpecies, which reduce under one head as many
of the génera as agree in any one circumftance; from which the fpecific charaćter is
formed. If the fpecimen examined have a
twining ftem it muft then be referred to the
firft divifion; if the peduncles are two-flowered,
to the fecond; if many-flowered, to the third.
But the génera muft be well underftood before any attempt is made to inveftigate the
fpecies; and when they are entered upon,
many obfervations may be found in the Génera Plantárum, noted beneath the géneric
charaćters, which may be very ufeful in elu-
cidating the fpecific diftinćtions. There is
another work of Linneus's, the Spccies Plantárum, which gives an account of the fpecies
only,

only, with their varieties. This work is not
tranflated, which is much to be lamented,
though the Syftem of Vegetables in part fup-
plies it's place, and is much to be preferred
to it, being an abftract both of the Species
and Genera Plantárum. The Syftem of Vegetables is a work of wonderful ingenuity;
there are to be found in many fingle pages of
it twenty plants accurately difcriminated from
every other known plant; and more than
10,000 plants are defcribed in the compafs of
one octavo volume. The tranflation of this
work cannot be too highly prized by all who
are unacquainted with the Latin language,
and are defirous of ftudying botany. The iris
is a flower liable to perplex the young botanift; but in obferving the fame order of
inveftigation as that recommended in the
canna and lonicéra he will readily be able to
refer it to it's genus. The character, " petal-
like," of the ftigma, diftinguifhes the iris
from feveral other génera of the clafs Triándria
and order Monogy'nia, with which it is arranged, although, before the flower is diffected,
the trifid divifions of it's fummit might be
miftaken for petals. The whole form of the
flower is beautiful, the corol is fix-parted,
the

the three outer divifions falling back, the three
inner erect, and all joined together by their
claws, the ftigma " petal-like." By ftripping
off the fix-parted corol the ftigma may be
plainly feen. Under each of it's three divifions is a ftamen preffed down upon the falling
petals of the corol. Some fpecies have a
beautiful fringe along the middle of thefe reflected petals, which is the nectary; others
have another kind of nectary, confifting of
three honey-bearing dots, externally, at the
bafe of the flower. The capfule alfo varies
in different fpecies; in fome it is three-
cornered, in others fix-cornered. Thefe are
obfervations on the family of the iris which
are very ufeful. Such génera as are nearly
allied to each other are placed in regular order; and if their affinity is great, the circumftance which feparates them into diftinct
families is noted.

The circumftances of colour, fmell, or
tafte, however effential to the ufe or agree-
ablenefs of the flower, are liable to vary fo
much, that they are by no means proper to
enter into either the géneric or fpecific characters of plants, which ought always to be
taken from fuch marks as are moft conftant.

On

On this account Linneus has rejected the dimenfions of the parts, except relatively, one to
the other ; place of growth alfo is too uncertain
to be admitted as a decided character: but all
thefe circumftances of fmell, tafte, colour,
fize, and fituation, are noted after the fpecific
characters in the Species Plantarum, and have
their ufe, if taken in aid of the more decided
marks of difcrimination. Linneus efteemed
the nectaries of greater importance in determining the génera, than almoft any other
part; and, by the ufe he has made of them,
has eftablifhed their confequence, although fo
much neglected and overlooked before his
time that they had not even a name. In the
clafs Monadélphia, one-brotherhood, the orders depend on the number of ftamens; and
the genera contained in thoie divifions are
again feparated by their number of piftils.
But although this is the leading character, it
is by no means fufficient to diftinguifh the
families from each other. The manner of
growth of the feeds, or the veffel by which
they are contained, with the number of divifions of the calyx, are frequently had recourfe
to in the difcriminations of the génera. From
the numerous kinds of geraniums, and the

K

variety

variety obferved in their number of ftamens,
Linneus found it neceffary to arrange them
under different heads, as may be feen in the
Syftem of Vegetables. Thefe divifions being
chiefly regulated by the vaiiation in the number of ftamens, could not but perplex the
young botanift, from being in direct contradiction to the character of the order under
which they were primarily affembled. L'Heritier's new arrangement of the geranium
tribe has removed thefe difficulties, and added
great improvement to the Monadélphia clafs.
He has divided the family into three diftinct
génera, Erodium, Pelargónium, and Geránium; the names Erodium and Pelargónium
fignifying heron's bill and ftork's bill, as Geranium fignifies crane's bill. Erodium includes Linneus's divifion with five perfect
anther-bearing ftamens; Pelargónium thofe
with feven anther-bearing ftamens; and Geránium thofe with ten. It is doubted whether
the genus Geránium may, with ftrict propriety, be claffed with the flowers of one-
brotherhood, as it has not it's ftamens decidedly
united at their bafe; at prefent it remains in
the clafs to which Linneus referred it, and
probably will be continued there, as the appearance

pearance of the ftamens and piftils fo much refemble thofe of all the one-brotherhood flowers,
that, without very nice examination, the
want of union at the bafe is not eafily difcovered. Four of our Britifh fpecies of geranium ought now to be arranged under the
genus Eródium, only five of their anthers
bearing ftamens; thefe are the cicufánium,
the pimpinellifólium, the mofchátum, and the
maritimum.

Dr. Smith, in his agreeable and ufeful publication of englifh botany, has thrown much
light upon the genus Geránium. He has
fhown us that the aril of the feeds varies fo
much in the different fpecies that a better
mark of diftinction cannot be had recourfe
to. His elegant and truly fcientific work
fhould be in the hands of all young botanifts
who are defirous of becoming acquainted
with the plants of their own country. In the
clafs Syngenéfia, united anthers, the form of
the corol of the feparate florets, or the manner
in which they are placed on their common
receptacle, are the marks by which the different orders are divided. By tracing fome
of the larger flowers to their génera the method of ftudying this intricate clafs will be

K 2   beft

beſt underſtood. When the pupil has provided himſelf with an artichoke (cy nara ſco-
lymus), he will find the florets of which it
conſiſts all of them to contain both ſtamens
and piſtils: this circumſtance refers it to the
firſt order. The firſt diviſion of that order
compriſes that ſpecies of corol termed, by
Linneus, ligulate, or tongued. The artichoke
cannot have a place among the flowers aſſem-
bled under this character, the corols all being
tubular. The next diviſion is marked by the
flowers being *headed*, the mode of infloreſcence
which is found in the plant under examination. In this diviſion are arranged ten génera.
The different characters of the firſt five by no
means agree with the artichoke; but the obvious marks of the " calyx ragged, with ſcales
channelled, thorny," refers it immediately to
the genus Cy'nara; and on examining the
more diffuſe deſcription at No. 928, there
can no longer remain a doubt that it is of
that family: the beautiful pappus which
crowns the ſeeds, and the ſize of the receptacle, which is the part we eat, are objects
well worthy of obſervation. In dandelion the
florets are all furniſhed with ſtamens and
piſtils, and of the ligulate form. In the

numerous

numerous génera comprifed under this head,
the receptacle is the firft mark of diftinction;
that part of fructification in the dandelion is
naked, or clear from either down or chaff;
the calyx is imbricated with loofe fcales; a
circumftance found in this genus only: the
plant, therefore, is leóntodon. There is, however, another character which ought to be
attended to.; this is the pappus. The diftinction betwixt. plumed and hairy may require
fome explanation. The pappus of feeds in
the compound flowers is either formed of
fimple hairs, or of hairs fet with other finer
hairs. In the former cafe the pappus is called
hairy; in the latter plumy, or feathery: the
pappus of artichoke (cy'nara) is hairy. In
the leóntodon the pappus, " plumy ftiped,"
or fixed upon a fhort foot-ftalk, is an effential
character of the genus; though, not being the
only one, is not of fo much confequence. In
dandelion (leóntodon taráxacum) this mark is
not found; and in the obfervations beneath
the generic characters, in the Génera Plantarum, this deficiency is remedied, and alfo
fome peculiarities in a few other fpecies,
which might have feparated them from their
genus with as much propriety as the taráxacum

K 3   has

has been removed. Tragopógon, goat's-beard, exhibits a specimen of the plumy pappus; in the artichoke this part is distinctly hairy. This minute circumstance respecting the pappus of seeds is of great use in marking the génera, therefore should be attended to: if it is exposed a little to the air to dry it will then be more clearly perceived of which kind the pappus may be esteemed. The deficiency of the plumy pappus in dandelion has been thought sufficient, by Scopoli, to make another genus of it, which he has named Hedypnois. However, as Linneus has uniformly shown his disapprobation of multiplying the génera from the single circumstance of an individual differing in any one part of fructification from it's family, it would, perhaps, be better to follow his method in this respect. There may be frequently found, in the compound flowers, distinctions obviously marked. In the burdock (arctium lappa) the outer scales of the calyx are hooked at the extremity with very sharp shining hooks. The onopórdon, cotton thistle, is distinguished from the cárduus, the true thistle, by having a receptacle somewnat like a honeycomb, that of cárduus being hairy; and hence may be per-

ceived

ceived the excellence of the Linnean method. Mr. Curtis has, in many génera of this difficult clafs, difcovered conftant marks by which they may be diftinguifhed in different ftates of growth. In the onopórdon acánthium, when the flowering is over, he has obferved that the innermoft fcales of the calyx clofe ftrongly together, and preferve the feed, contrary to the calyx of cárduus, and moft other génera of the compound flowers, which, as has been before remarked, expand and difperfe their feeds. The fmaller flowers of this clafs are more difficult to inveftigate; but, if proceeded with in the fame manner as the larger kinds, a competent knowledge of them may foon be obtained. A numerous tribe of plants, termed the umbelled plants, which are contained under the clafs Pentándria, will be found more eafy of accefs to the young botanift if he has fome previous information in the mode of their inveftigation. The umbelliferous plants fhould be gathered for examination before their florets are wholly expanded, otherwife it will not be eafy to determine the clafs to which they belong, as the anthers frequently drop off as foon as they arrive at maturity. If this is attended to, it

K 4 will

will not be difficult to trace their characters
of both clafs and order, Pentándria Digy'nia.
Under this order are comprifed feven divifions. The umbelled tribe are collected under the character of their mode of inflorefcence,
their florets having " five petals, *above*, and
two-feeded." This divifion is again feparated
into three parts, the firft diftinguifhed by the
flower having an univerfal and partial involucre; that is, each collection of florets being
furnifhed with an involucre, and all together
being contained by one at their bafe; fecond,
with partial involucres, and no univerfal one;
and the third, without involucre, either univerfal or partial. In the inveftigation of the
further generic characters the pupil may be
fomewhat perplexed by the fimilarity of terms
ufed in the diftinction of umbel-bearing plants
and thofe of the clafs Syngenéfia. In this
clafs, which confifts of the compounded
flowers, the term radiate is applied to thofe
génera which have their florets of the circumference flat, and thofe of the centre tubular.
In the umbellate tribe of plants the term radiate is made ufe of to diftinguifh the umbels
which have the flowers of the circumference
of a larger fize than thofe of the centre; in

of

which cafe it frequently happens that fome of the florets are deficient in either the fta-mens or piftils, and thence do not all produce feeds; from which circumftance Linneus has termed them abortive, as he has called thofe umbels fertile, the florets of which are all productive of feeds. The term flofculous, made ufe of in defcribing the compound flowers, marks thofe that have all their florets tubular, applied to the umbelled plants of Pentándria Digy'nia. It fignifies thofe um-bels, the florets of which are all of the fame fize. The term uniform is made ufe of in the Génera Plantárum to mark thofe flowers which are called flofculous in the Syftem of Vegetables. Not uniform is applied to thofe termed radiate. The form of the feeds is alfo a circumftance to be attended to in the difcrimination of the fpecies of thefe flowers; and both feeds and flowers may generally be found at the fame time in a proper ftate for inveftigation. The fcandix pecten, fhep-herd's needle, is diftinguifhed by the very long beak with which the feeds are fur-nifhed. A fpecimen of the radiate flowers may be feen in this genus. the florets of the

dilk

difk being often male, or containing only ftamens. The difk and ray are the terms made ufe of to exprefs the centre and circumference, and are frequently applied, with the fame meaning, to the compound flowers. In the fimple flowers of the clafs Pentándria there are fome génera the fpecies of which differ fo much in fome parts of their fructification, that it may be neceffary to apprize the young botanift of this diffimilarity. The gentianella and leffer centaury, both placed by Linneus under the genus Gentiána, are fo unlike in their appearance as even to perplex an experienced botanic eye. The ftructure of thofe fpecies of Gentiána, which are known by the name of Gentianella, is fo peculiar as to feem to give them a right to form a feparate genus; and the centaury is now placed by Mr. Curtis in the genus Chirónia, from the circumftance of the anthers becoming twifted after they have fhed their duft, a diftinguifhing character of the Chirónia genus, alfo from the fimilarity of their outward habits. Such refpectable authority as that of Mr. Curtis muft have great weight; and all who underftand the

value

value of the works of Linneus muſt acknow-
ledge with gratitude the advantage they have
derived from the labours and candid criti-
ciſms of that much-lamented and accurate
botaniſt.

LECTURE

LECTURE II.

Nectaries of Plants.

THERE are some very common plants which, either from the natural structure of their fructification, or from some adventitious circumstance, are not easy of investigation to the young student. The house-leek (sempervivum tectórum), a plant of the class and order Dodecándria Dodecagynia, twelve stamens, twelve pistils, is subject to so extraordinary a change in it's parts of fructification as might nearly baffle an experienced botanist in the inquiry after it's genus. This perplexing appearance is accurately described by Mr. Curtis from Haller, who has given a very minute account of this plant. It's filaments frequently, even while young, are evidently enlarged towards their ends, and throw out from their substance little oblong white corpuscles, like the eggs of some insect: the filaments thus enlarged, are more glutinous than those in their natural state, and have

their

their anthers fomewhat imperfect. As the
fructification advances towards maturity, the
filaments continue to enlarge about the middle, while the top is drawn out to a kind of
beak, in which ftate they might be miftaken
for the piftil. On cutting them through they
appear hollow, and to contain fome of the
fame corpufcles, which may be feen on the
outfide of many of them, fo that it would
be impoffible to know them to have been
originally filaments. This fhows you the advantage of examining flowers in their different ftates of maturity, and before the full
expanfion of their corols. The fempervivum
is nearly allied to the fédum, but differs in
having more than five petals; it is alfo liable
to increafe in it's number of piftils, when it
grows luxuriant.

We are obliged to Mr. Curtis for an accurate knowledge of the difficult and curious
genus Euphórbia, which is the botanic name
of the churn-ftaff. He juftly remarks, that the
Linnean characters of this family will not, in
any of the Britifh fpecies, even guide us to it's
clafs. The ftamens are very minute; there are
feldom more than two or three that appear
above the calyx, the reft are concealed within

3   it,

it, and rarely amount to twelve in number, so that it fails in the eſſential character of the eleventh claſs, wherein it is placed, that character requiring that the flowers contained in it ſhould not have fewer than eleven ſtamens, or more than nineteen: the ſmallneſs of the ſtamens, and the milky juice, which flows ſo plentifully from every part when bruiſed, renders the inveſtigation of the Euphórbias, on the principles of the Linnean ſyſtem, extremely difficult. A clear idea of the flower and fruit of this ſingular genus may, however, be obtained by diſſecting ſome flowers of the large garden ſpurge-tree, or euphórbia láthyris. The part which Linneus had called the corol, Mr. Curtis has now named the nectary. There is a ſingular appearance which crowns the ſeeds of theſe plants, and which did not eſcape the notice of Mr. Curtis. This extraordinary appendage is termed by him a button: it is of a fleſhy ſubſtance, of a grayiſh colour, heart-ſhaped, and ſtands looſely on a ſhortiſh foot-ſtalk. In the tree-ſpurge it gives beauty to the large black ſeed which it crowns. The outer habits and milky juices of the euphórbias are ſufficient marks of diſtinction of this genus;

but

but the curious ſtructure of their fructification well repays the trouble of the moſt minute inveſtigation.

We now proceed to the Nectary, which has been defined by Linneus to be that part of the corol which contains the honey, having a wonderful variety both as to ſhape and ſituation, ſometimes being united with the petals, and ſometimes ſeparated from them. The lower part, or tube, of one-petalled corols, generally is found to contain a ſweet juice, which is the honey. In the flowers of árbutus unédo (ſtrawberry-tree) it is ſo profuſe as to run out, when the corol is opened, and to give the flowers a ſtrong ſcent, reſembling that of the honey of bees; it is alſo found at the baſe of the petals, in many of the butterfly tribe of plants. Clover (trifólium praténſe) contains much of this liquid. The chief diſtinctions of the nectaries, which adhere to any of the parts of fructification, are, *firſt*, the ſpur-form, which is found in one-petalled flowers, as ſnapdragon (antirrhínum), and valerian (valeriána); and in many-petalled flowers, as in órchis, lark-ſpur (delphínium), and víola. *Second*, ſuch as are on the inſide of the petals,

7   as

as in crown-imperial, and all the family of fritillária, though in none fo obvious as in the fpecies imperiális, in ranúnculus, and dog tooth (erythrónium): the nectary in lily (lílium) is that raifed line which runs down the petal lengthways. *Third*, the nectaries which crown the corol, as in paffion-flower (paffiflóra), narcíffus (ly'chnis). *Fourth*, on the calyx, as in nafturtion (tropæ'olum), being a fpur attached to the calyx. *Fifth*, on the ftamens, which in bay (laúrus nóbilis) are three glands ending in two briftles, furrounding the germe. *Sixth*, on the germe, as in fome fpecies of iris, and in hyacinth, and the plants of the claf: four-powers, Tetradynámia. *Seventh*, on the receptacle in fempervívum, and mercury (mercuriális).. *Eighth*, all thofe nectaries which are not apart from the corol, but the fingular conftruction of which does not admit of their being placed among any of the kinds I have enumerated, as in nettle (urtíca), the nectary is fituated in the centre of the ftamen bearing flower, very fmall, in the form of a cup. In fact, the term nectary is applied by Linneus to every part of fructification, which, from it's fingularity, cannot be ranked among the feven

regular

regular parts of a flower. It has been doubted whether this part exifts in every flower, and certainly we find many deftitute of it, as a diftinct apparatus; but if any part, wherein this fweet juice, called honey, is found, has a right to be termed a nectary, it may be decided, that there is no flower without it; and that Linneus was of this opinion appears from his having named it, in the Syftem of Vegetables, as a conftant appendage of the corol, calling it the honey-bearing part proper to the flower, diftinguifhing it into two kinds, *proper*, when diftinct from the petals and other parts, *on the petals*, when forming a part of the corol. It's not being noticed in many of the génera may feem an objection to Linneus having confidered it as a conftant part of the fructification; but he could not be ignorant of it's exiftence in the compound flowers, the lower part of the florets, of which they confift, generally containing the juice in queftion, and yet he has not named it in any of the génera of the clafs united anthers (Syngenéfia), except thofe of the order Monogamia, or fimple flowers, which have fpurform nectaries; whence we may conclude he omitted it in all thofe génera, where it's

L   ftructure

ſtructure was not ſuch as to form a marked
character. As a further proof of this, the nectary is not named in the one-petalled flowers,
though nothing can be more evident than the
honey contained in their tubes; and Linneus
has, in ſome of his works, called the tube of
a one-petalled corol a true nectary. Among
the nectar-bearing ſtamens he enumerates
thoſe of the fraxinella (dictámnus). It ſeems,
however, more probable, that the reſinous
matter, with which they abound, is not of
the nature of honey, but ſimilar to that found
upon the ſtalks, which is ſo inflammable as
to take fire on the approach of a lighted candle, and to burn like ſpirit of wine, till it is
entirely exhauſted.

The ſtructure of thoſe nectaries which are
placed ſeparate from all other parts of the fructification, is an object that merits the ſtricteſt
attention, not only as diſtinguiſhing decidedly
one genus from another, but from the artful
manner in which they are formed for the
purpoſe of preſerving from inſects the precious ſtore contained in them. The moſt
remarkable are thoſe of the monk's-hood
(aconítum napéllus), of chriſtmas roſe (helléborus niger), parnáſſia, and columbine (aquilégia),

légia), and of the órchis tribe. In aquilégià
the nectaries have been thought to refemble
the neck and body of a bird, and the two
petals ftanding upon each fide to reprefent
wings, whence it's name of columbine, as if
refembling a neft of young pigeons, while
their parent feeds them. In helléborus the nectaries are placed in a circle like little pitchers,
and add much to the beauty of the flower;
but there are not any which are a greater ornament to the flower than thofe of the parnáffia.
The beautiful tranfparent globules which
fringe the margins of the five fcales, called
nectaries, may probably contain fome vifcous
juice, which ferves to guard the honey from
the depredations of infects. In the careful
diffection of a pink, when the ftamens firft
become mature, the bafe of the calyx will be
found replete with honey. By what part of
the fructification this juice is fecreted, is perhaps not an eafy matter to determine; but if
determined, that part muft undoubtedly be
termed the nectary. The nectaries of the
flowers of mignonette (reféda odoráta) are
of curious and elegant conftruction, two
fringed petals growing clofe together form a
little cafket, or box, the lid of which is a

L 2 fmall

small scale growing betwixt the stamens and petals, and pressing so closely on the latter as to shut up securely a small drop of honey in the hollow formed by their union; and bees may be frequently seen baffled in their attempts to plunder this honey, not being able to open the lid sufficiently wide to allow of the insertion of their trunks. The curious structure of the genus Passiflóra merits minute examination. In the common passion-flower the large size of the parts of fructification renders the examination of the position of the stamens and pistils peculiarly easy. The petals and calyx nearly resemble each other in front, both being of the same form and colour; these beautiful rays are the nectaries; the stamens are five, having at the first view, the appearance of being placed on the pistil, but in reality growing from the bottom of the germe, where it joins the little pillar on which it is elevated. The three large styles are very evident, and, from their purple colour, and that of their stigmas, give much beauty to the flower. The nectaries form the principal feature in the flowers of this genus, and in some of the species have the appearance of a basket made of

blue

blue and white beads ſtrung upon wire. The generic characters of paſſiflóra, given by Linneus, do not agree with many of the ſpecies; and it admits of ſome doubt whether the ſtamens can be properly ſaid to grow on the germe. Perhaps the ſmall pillar, to which both the ſtamens and germe adhere, might, with more propriety, be eſteemed a receptacle. Linneus calls this pillar a ſtyle; but, if it be one, we are at a loſs to know what part of the flower theſe three apparent ſtyles, with their ſtigmas, muſt be called, and to which he alſo gives the name of ſtyles. This is one of the few génera that we find not juſtly deſcribed.

It is not an eaſy matter to obtain a diſtinct idea of the parts of fructification of the órchis tribe: a peculiarity of ſtructure runs through the whole of them, ſo different from what we commonly meet with in other plants, as to make them well worth inveſtigating. I have given, in Plate the Firſt of the Second Part of this Work, an engraving of a ſingle flower of the early ſpotted órchis on it's peduncle, with it's bract or floral leaf, in which may be ſeen the twiſted germe, the petals, the lip, and form of the nectary, of their

L 3   natural

natural fize. I have alfo given an engraving of the feparate parts magnified: with thefe the natural flower fhould be compared. Each flower contains two ftamens, the ftructure of which is very curious. Each of thefe ftamens is contained within a bag or cafe, the edges of which fold over each other, and open in front, as the plant advances towards maturity. At this period, in many of the órchis tribe, they hang down, out of their cafes, towards the ftigma, and on the flighteft pull they are drawn out. If gently drawn with a fine needle, they will be found elaftic; and a fmall tranfparent globule may be feen at the bafe of each fta-men, and at the top a club-fhaped fubftance, in moft of the fpecies of a yellow colour, the furface of which is covered with fmall grains; thefe muft be efteemed anthers. In a mag-nified view of the ftamens the anthers will be found compofed of irregularly fquare cor-pufcles united together by fine elaftic threads. That thefe corpufcles produce the fame effect as the anther duft of common flowers, feems highly probable, although, at prefent, the manner of their doing fo is not known.

Many of the órchis tribe have their feed-veffels large, well formed, and filled with feeds,

feeds, which, though extremely minute, appear perfect. The fmallnefs of the feed is certainly no argument againft it's power of vegetating. Some of the ferns, the feeds of which are much fmaller, are well known to be propagated from feed, and to come up fpontaneoufly in hot-houfes, where the original plant has fcattered it's feed; and probably by minute attention the feedlings of órchis may be difcovered. However, I am of opinion, that the órchifes are propagated from feed, as many young plants of them are frequently found together, and it is well known that they never increafe plentifully by the root; but in this, and all other parts of natural hiftory, we can only hope for fatisfaction from accurate and repeated obfervation. The art of making experiments is, however, poffeffed by few, and requires much patience, added to an accurate and impartial judgment. If we watch a bed of órchifes, in the hope of finding feedlings on it, we fhall eagerly catch at every circumftance that can favour this hope. It is the bufinefs of an experiment maker to be always looking for circumftances which make againft his theory, and not for it; and to ftate as ftrongly what he remarks

L 4

unfavour-

unfavourable, as favourable to his wishes.
The early spotted órchis is easily distinguished
from every other known species; it's spotted
leaves and large bright purple flowers will
generally be marks sufficient; but should the
young botanist please himself with the suppo-
sition of having gathered a variety of kinds of
órchis morio, he would be much disappointed
to find, on examination, that they belonged
to one species only; an instance which shows
how little to be relied on are the colours of
the corol, which in this species assumes all
changes of colour, from a deep purple to a
white. Yet, under all it's varieties, this flower
is distinguished from all other british órchises
by retaining more or less strongly the character of having it's two outermost petals
marked with green parallel lines. In this
órchis the anthers are of a green colour.

There are ten distinct british species of the
real órchis; but by common observers some
other génera have been confounded with them,
which ought not to have been so. Linneus has
distinguished the different génera of these curious plants by the form of their nectaries. The
flower commonly known by the name of
bee órchis belongs to the genus of óphrys, and

3

is

is the fpecies apifera, bee-bearing. The dif-
tinguiſhing character of óphrys is the nectary
hanging down longer than the petals, and being flightly keeled behind only. That fpecies,
commonly called the tway-blade, is the egged
ophrys. By comparing thefe flowers with the
plates of Mr. Curtis's London Flora * they will
be found moft accurately given; and the
great difference in the ftructure of the órchis
and óphrys génera will be well feen. Thefe
génera are alfo greatly elucidated by the obfervations of Dr. Smith in his Englifh Botany.
Linneus has formed the fpecific characters of
feveral of thefe flowers from peculiar circumftances found in the nectary; that of the
tway-blade, or óphrys ováta, is marked by
it's nectary being two-cleft. The leaves of
thefe two fpecies of óphrys differ materially
from thofe of the órchis tribe. The root of
the óphrys apífera refembles thofe of the
órchis genus, which are bulbous, but that of
the ováta is fibrous. Linneus, in the generic
characters of the four families of órchis, fa-
ty'rium, óphrys, and ferápias, which are all

* For the convenience of thofe, who may not have accefs
to that valuable publication, a plate of the órchis and óphrys
is given at the end of this Lecture.

clofely

clofely allied, marks the circumftance of the germe being twifted as a peculiarity common to them all. It certainly does not run through all the fpecies, and might probably be found exclufively to belong to the órchis genus.

EXPLANATION

EXPLANATION OF PLATE I. PART II.

PARTS OF FRUCTIFICATION OF HIPPURIS, CANNA, EUPHORBIA, ORCHIS AND ARUM, AND THE NECTARIES OF PARNASSIA AND ACONITUM NAPELLUS.

Fig. 1. Part of a Spike of Hippúris Vulgáris, with the flowers in the bofom of the leaves, a.

Fig. 2. A Flower of Hippúris Vulgáris magnified.

Eig. 3. Anther-bearing Petal of Cánna, b. With the Style growing to the Petal-form Filament, c. d, The Stigma.

Fig. 4. Three-leaved Perianth of Cánna growing upon the Germe.

Fig. 5. A Flower of Euphórbia Heliófcopia magnified. e, The Calyx. f, The Nectary. g, The Stamens. h, The Germe. i, The Stigma.

Fig. 6. Seeds of Euphórbia to fhow the fmall white button at the upper end, k.

Fig. 7. Nectaries of Parnáffia and Aconítum Napéllus, Monk's-hcod.

Fig. 8. Stamens and Stigma of Paffion Flower.

Fig. 9. An entire Flower of early fpotted Orchis. l, The Bract. m and n, The Petals. o and p, The lip and horn of the Nectary. q, The twifted Germe.

Fig. 10. The Stamens magnified. r, The Glands at their bafe.

Fig. 11. A Stamen magnified with the Anther drawn out.

Fig. 12. A Flower of Ophrys Ováta. s, The Cloven Nectary.

Fig. 13. A Flower of Ophrys Apífera, Bee-ophrys. t, The Petals. u, The Nectary.

Fig. 14. A Flower of common Arum. v, The Anthers. w, The Germe. x, The Nectaries above and below the Anthers.

Plate 1. *Part II. P.157.*

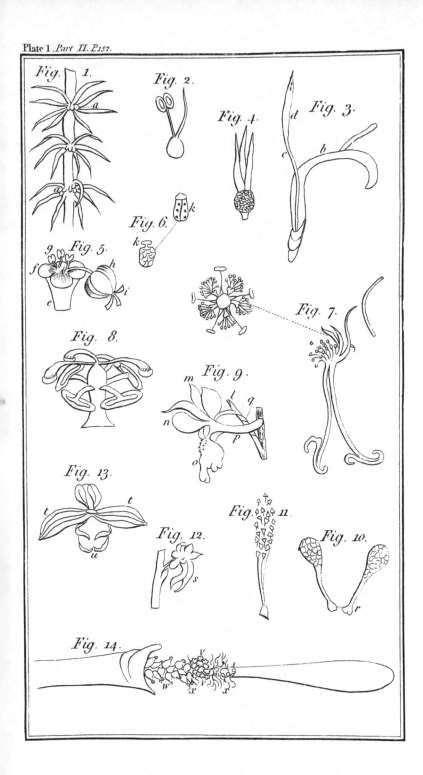

London, Published May 1.*1797* by J.Johnson S.*t* Pauls Church Yard.

LECTURE III.

Inveſtigation of different Génera of the Claſſes One-houſe and Two-houſes. Of Ferns.

HOWEVER extraordinary the ſtructure of the génera juſt now conſidered may appear, there is yet another genus of the claſs Gynándria which, in the curious mode of it's fructification, ſurpaſſes them all; this is the árum, of which the britiſh plant known by the common name of wake-robin, is a ſpecies. This plant is ſubject to great variety in it's colours. The part called by children the tongue varies from a yellowiſh green to a deep purple; the leaves and heads differ in ſometimes being beautifully ſpotted with black, at others plain green; the leaves alſo are found of different ſhapes. This is a won derful flower, and ſeems intended by nature to ſhow us, that ſhe is not confined to any one method of renewing her productions. Here are berries produced with perfect ſeeds, which

germinate

germinate and continue the species, as certainly as those seeds formed in plants, which
we call of a more natural structure, because
they are of one more common. All other
known plants have their pistils placed within
the stamens. In the árum the stamens are situated rather more inward than the pistils, and
above them on the receptacle. These stamens
are not raised by filaments, but are a collection
of anthers four-cornered, and growing to the
club-form receptacle; above and below these
anthers are placed several roundish bodies, terminated by a tapering thread; these Linneus
calls the nectaries. Beneath the lower order
of nectaries, the seed-buds are placed, sur-
rounding the base of the spadix, or tongue,
of an oval shape, without styles, and their
stigmas bearded with soft hairs. These seed
buds become berries of a beautiful bright
scarlet colour, corresponding in number with
the germes; are round, and have one cavity.
The younger Linneus was of opinion, that
the árum did not properly belong to the class
Gynándria, but that it should be placed in
the class One-house, as every anther and
stigma were rather to be esteemed distinct
florets, than as belonging to one common
flower;

flower; at prefent it remains in the clafs Gynándria. The root of this árum is extremely
acrid; but that property does not prevent it's
being dug up and eaten by the thrufhes. Some
fpecies have their roots fo mild as to make a
part of the food of the inhabitants of the
hot countries, where they grow; and fome
of the forts are cultivated by the inhabitants
of the South-Sea ifles, and of the fugar colo-
nies, as efculent plants. The leaves of one
of the fpecies, called indian cale, are boiled
to fupply the want of other greens. The
roots of the árum maculatum, which is the
Britifh fpecies, were formerly ufed for ftarch;
Gerrard mentions it having been fo, and adds,
that it was fo extremely acrid, that the people
who made ufe of it had their hands fo much
chapped, that they were healed with difficulty.
This property is not alone confined to the
root, the whole plant abounds with an acrid
juice.

Much curiofity and beauty of ftructure are
to be found in the flowers of a genus of the
clafs Diœcia, hydrócharis, or frog's-bit. This
plant is of aquatic growth, and one of the
moft ornamental of our water plants. The
leaves, the whole ftructure and economy of
this

this plant, are exceedingly curious, and merit minute examination. The male flowers of the hydrócharis have nine ftamens, difpofed in three rows. The filaments of the middlemoft row put out from their bafe, on the infide, a ftyle-like fubftance, which is placed in the centre of the flower. The two other rows are con-nected at the bottom, fo that the internal and external filaments adhere together. The anthers are yellow, nearly linear, and have two cavities. Linneus does not take notice of the nectary, but Mr. Curtis has obferved, in the female flower, three yellow glands crowning the germe, to which he gives that name. The fpathes of the flowers give the plant fomewhat the appearance of fea-wrack (fúcus). Thefe buds, from their tranfpa-rency, have the appearance of bubbles, and are very numerous, both in the male and fe-male plants, and chiefly grow near the root. In the male there are alfo a pair of thefe fpathes, which grow out about the middle of the flower-ftalk, and look like little bladders, containing the tender unopened flowers. Mr. Curtis·differs from Linneus in defcribing the female flowers as enclofed by a fpathe, which contains only one flower, that of the

male

male three or four. Among the aquatic
plants we find not only beauty but mag-
nificence; the greater and leffer ty pha, with
their yellow downy fpikes, attract the eye of
the botanift from a confiderable diftance, but
are not fatisfactory to a novice in the fcience.
Their fiowers, confifting of very minute
parts, are difficult of inveftigation; Mr. Curtis's account of them fomewhat differs from
that of Linneus, and is to be preferred; as
he examined all the parts accurately with a
microfcope. Thefe plants are of the One-
houfe clafs, and by Linneus are placed in the
order three-ftamens; but as on one filament
are found one, two, three, or four anthers, it
feems that they might more properly have
been arranged in that of Polyándria, or many-
ftamens. What Linneus has called the calyx, from Mr. Curtis's obfervations, does not
appear to be one, but rather fome hairs
proceeding from the receptacle, which is covered by them after the ftamens are fallen
off. Thefe fpikes of flowers are aments, or
catkins, and their cylindric form marks the
effential character of the genus The male
flowers are numerous, and terminate the
culm, which is the term that Linneus gives

M   to

to the ftraw of graffes, and the reed-like
plants. The female flowers are alfo numerous, and entirely furround the culm. The
ty'pha major, when it's fpike of ftamens is
nearly ripe, makes a magnificent appearance;
indeed, every part of this plant deferves attention: the root derives much beauty from it's
fine mofs-like fibres, and the fhades of brown
and green, with which the upper furface is
varied.

The numerous genus cárex, in clafs Monœcia, one-houfe, may perplex a young botanift in the mode of their inveftigation, their
flowers being fmall, and growing clofely together; but, if each feparate floret be examined before the anthers are arrived at
maturity, their genus may be more eafily de-
tected than from their firft appearance might
be fuppofed. Particular attention fhould be
paid to the ftate of the ftamens in all plants
of the catkin, or ament, kind; and if that circumftance is regarded they will not be found
difficult of accefs. Some of the fpecies of
cárex are obvioufly diftinguifhed by their outward habits. The cárex pendula, in whatever fituation it is found, is diftinctly marked
by it's long pendant female fpikes. Thefe
are

are very slender while young, but become much thicker as the seeds ripen. It's fructification merits examination, as indeed does that of the catkin tribe in general.

It is necessary for the pupil to obtain some idea of the structure of the Cryptogámia plants; he should therefore begin with the ferns (fílices). The plants contained in the class Cryptogámia have not yet been observed to bear either stamens or pistils; therefore, when the term fructification is applied to them, it has no farther signification than the seed, and the apparatus by which that is contained and dispersed. The whole tribe of the filices, or ferns, is divided into three sections, from the manner in which their fructifications are disposed. The first division consists of such as have their fruit in spikes; the second, of those which have it placed on the under side of their leaves; and the third, of what is termed by Linneus radical fructification; a specimen of which is well seen in the pepper grass (pilulária). The botanical world is much indebted to the accurate researches of the celebrated Hedwig for many important discoveries in the obscure families of plants belonging to Cryptogámia. Of the

spiked

spiked fructification a better specimen cannot be examined than the equisétum sylvaticum, at the time when it is beginning to disperse it's seeds; in the progress of which there may be observed appearances which seem to have a right to be considered as stamens and pistils. In the investigation of this plant recourse must be had to glasses; but it will be found more agreeable to view the parts through a microscope when some idea· is obtained of their structure from engravings; and I recommend to the student, when obliged to have recourse to plates, to remember that he there relies on the authority of others; whereas in botany, as in all other things, small progress can be made if he does not take the trouble of seeing for himself. It is the observance of the rule; " See for yourself," that has rendered the works of Mr. Curtis so peculiarly valuable. Most of our botanical publications are taken one from the other: and thus, if an eminent botanist has, in the course of his researches, fallen into a mistake, the errour has been propagated. Mr. Curtis, from his caution in this particular, has done more towards the improvement of the science, than any other writer with whom I am acquainted;
and,

and, by his judicious and candid correction of the few errours in the works of Linneus, has rendered effential fervice to the botanical world.

But to return to the equifetum. Early in the fpring this plant puſhes out of the earth a little club-ſhaped head; round this head are placed, in circles, target-form ſubſtances, each ſupported on a pedicle, and compreſſed into angles, in confequence of their reſting againſt each other before the ſpike expands. Beneath each of thefe targets are from four to feven conical ſubſtances, with their points leaning a little inwards towards the pedicle. They open on the inner ſide, and on ſhaking them over a piece of paper, a greeniſh powdery mafs falls out, which at firſt is full of motion, but foon after looks like cotton or tow. All this may be feen without a microfcope; but by the affiſtance of glaffes green oval bodies have been difcovered, and attached to them (generally) four pellucid and very ſlender threads, ſpoon-form, at their ends, as may be feen in Plate the Third. Thefe ſmall woolley ſubſtances have, to the naked eye, no appearance of diſtinct formation; but we may always be fure, that a nice and regular

M 3 organ-

organization exifts in all the various parts of plants, though from the want of a proper method of inveftigating them this may not be always vifible to us. Thefe pellucid threads are almoft conftantly in motion, and are faid to contract themfelves upon the leaft breath of moift air, and, when wet with water, to roll round the green oval bodies from which they proceed. To fee this requires more powerful magnifying glaffes, and greater fkill in the conduct of them, than may probably fall to the fhare of botanifts in general; it will be well, therefore, at prefent, to take this curious hiftory upon truft: but an outline of the difcoveries of the moft eminent botanifts of our time ought to be known to all. Hedwig makes no doubt that thefe green oval bodies are the feeds, as they gradually increafe in bulk, and when they fall the fpike fhrivels; that the projecting fpikes are the ftigmas, and the conical fubftances under the targets are the capfules, and the pellucid threads, with the fpoon-form fubftances attached to them, the filaments and ftamens; the feeds are numerous, egg-form, or globular, placed upon and lapped up within the filaments of the ftamens. Future obfervations muft confirm or refute this opinion.

nion. The different appearance of the fuppofed feeds, with their ftamens, before the
burfting of the anthers and afterwards, feems
to be ftrongly in it's favour. The fcales, or
ftipules, which furround the flowering-ftalk
at certain diftances after it's protrufion, ferved,
whilft it was young, as a general fence to the
fpikes. From the inveftigation of the equifétum a clear idea muft be gained of the form
in which it's fructification appears, and
thence of that which may be found in the
reft of the génera, which are arranged in the
fpiked divifion of ferns. We now come to
that which contains the leafy fructifications,
the elegant conftruction of which cannot fail
to attract attention. The maiden-hair, a native of England, with it's purple ftalks and
fcolloped green leaves, dotted underneath
with innumerable fmall brown fpots, affords
a beautiful fpecimen of this curious mode of
inflorefcence. The fyrup of capillaire derives
it's name from the botanical appellation of
this little plant, capíllus véneris, and is fuppofed to be, in part, compofed of it; the minutenefs of it's parts renders them lefs proper
for examination than thofe of the larger fpecies of fern. The hart's-tongue (afplénium

fcolo-

fcolopéndrium), from it's fize, will fhow the
fruƈtification more diftinƈtly; the firft appearances of which, that can be obferved, are
fome little bags, or cafes of a yellowifh or
whitifh green colour, placed in rows on the
under fide of the leaves; if thefe are opened,
almoft as foon as they become vifible, there
will be found capfules, or feed-veffels, very
numerous, ftanding upright, and clofe together. At this time they appear to be of a
green colour; as they approach towards maturity, they change this for a dark brown;
at which period the cafes open lengthways
in the middle, and by the protrufion of the
capfules, the two fides are turned quite back,
and wholly difappear; this membranous fubftance may be confidered as the fame with
the calyx in other plants, and ferves to defend the tender capfules with their feed till
ripe, when their curious mechanifm ftrikes us
with grateful aftonifhment at the benevolent
and adequate care that nature takes of the
minuteft of her works. Each capfulé confifts
of three parts, the foot-ftalk, which fupports and conneƈts them to the leaf*; the

* See Plate Third of the Second Part.

jointed

jointed fpring, which nearly furrounds the third part; or cavity containing the feeds. The feeds being ripe, this cavity is forced open by the elafticity of the jointed fpring, and the feeds fcattered and thrown to a confiderable diftance, one half of the cavity remaining connected to one end of the fpring, and the other half to the other end. Thefe capfules are an agreeable fubject for the microfcope; but it is difficult to manage them fo as to gain a diftinct idea of their progrefs. They are placed fo clofely together on the leaf, that it is neceffary to feparate them from it with a fine knife, before they can be diftinctly feen. The warmth of the breath alfo, by occafioning the capfules to open and difcharge their feeds, gives them the appearance of fomething alive. While we are intently looking at one, hoping to obferve the operation, the ftrength and elafticity of the fpring, at the moment of difcharging, will often carry it out of fight; fo that to fee the manner of opening requires fome dexterous management, and much patience.

The roots of fome fpecies of fern have the appearance of different kinds of animals; that

that of the polypódium vulgáre as nearly refembles one of the very large kind of caterpillars, as the root of the polypódium
bárometz, if we may judge from the prints
of it, does a fheep! This plant is defcribed
by many eminent botanifts, as being deficient
in the elaftic ring, which furrounds the
capfules, and by means of which they are
burft open, and their feeds difcharged. It
would be extraordinary to find any of the
fern tribe deftitute of this feemingly effential
part; neither has it yet been difcovered, that
they are fo, by the accurate and diligent refearches of Mr. Curtis, who afcribes this errour
of defcription to the blindly following the
authority of figures; for had thofe authors,
who have falfely charaƈterized the polypódium vulgáre, from it's want of the elaftic
ring, made ufe of their own eyes, affifted
only by a common magnifier, they muft
have feen, what had long before their time
attraƈted the notice of inquiring botanifts.
At the fame time it is not eafy to account
for the errour of the ingenious Tournefort,
who has delineated the capfules of the genus
polypódium without rings; but this is one
of the many inftances which ought to deter
us

us from relying upon authority, be it ever
so respectable. There is one circumstance
attending this polypódium which does not
run through the whole genus, that is, the
want of the membrane, which, in the rest
of the family, is found enclosing the capsules: of this, however, it may not be destitute, but it may have escaped notice from
early falling off, when the capsules are arrived at a certain degree of maturity. This
tribe of plants not having been much attended
to leaves to modern botanists an ample field
of discovery; and the whole class Cryptogámia
is now become so much an object of inquiry
to persons of the first ability in the science,
that a few years will probably elucidate that
obscurity which has hitherto rendered it a
disgrace to Botany.

Having obtained a tolerably clear idea of
the fructification of ferns, practice and attention can alone render the pupil familiar with
the different génera; an undertaking in which
he will find much difficulty. So great a similarity runs through the fructifications of them
all, that the distinction cannot be founded on
that part of the plant. The various modes,
in which the capsules are placed on the frond,

3   or

or leaf, in fome of them, are ftrikingly different, and appear to form very diftinct and
fatisfactory characters; but when, as a tribe,
they come to be more minutely inveftigated,
the chracters of one genus are frequently loft
in thofe of another, and we in vain feek for a
precife generic character. The plates and
remarks in Mr. Curtis's London Flora are
particularly pleafing and ufeful on this fubject.
The elegance of the figures of fome of the
génera is fcarcely exceeded by their natural
appearance. Wherever the ferns are found,
they are ornamental; on walls, old wells, and
banks, in winter, they make a principal feature in that beautiful affemblage of the Cryptogámia plants, which may be faid to form a
winter garden.

EXPLANATION

EXPLANATION OF PLATE II. PART II.

HYDROCHARIS MORSUS-RANÆ, FROGS-BIT.

Fig. 1. A Plant of Hydrócharis Morſus-ranæ, Frogs-bit, to
ſhow it's outer habits and mode of growing. *a, b,*
Tranſparent Sheaths, containing Flower-buds.

Fig. 2. A Female Flower with the Germe, *c.*

Hydrócharis Morſus-ranæ.

Plate 2. *Part II. P.174.*

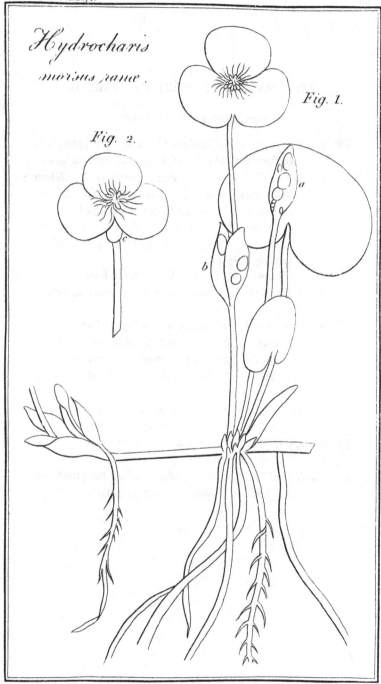

Hydrocharis

morsus ranæ.

Fig. 1.

Fig. 2.

London, Published May 1.1797. by J. Johnson, St. Pauls Church Yard.

EXPLANATION OF PLATE III. PART II,

Plate 3. *Part II.P.276.*

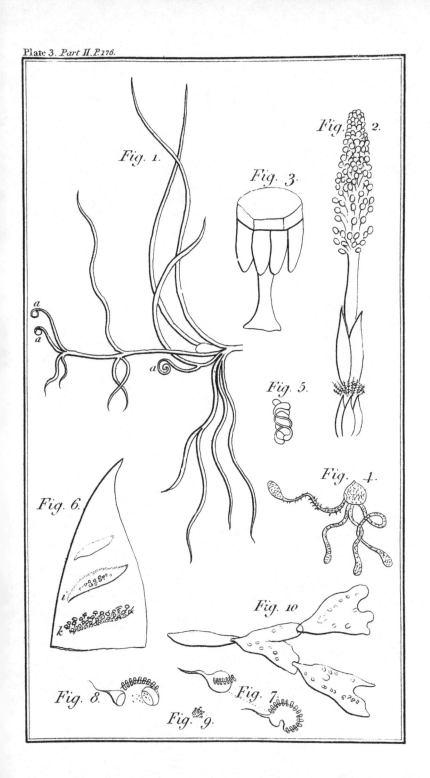

London, Published May 1.1797, by J.Johnson. S.Pauls Church Yard.

LECTURE IV.

*On the Moſſes, Flags, and Funguſes. Muſci, Algæ,
and Fungi.*

It is difficult to decide whether the palm
of beauty ſhould be given to the tribe of the
ferns or the moſſes; but from the extenſive
utility of the latter in the vegetable kingdom
they lay a ſuperior claim to our reſpect and
attention. The beauty of their leaves is too
obvious to require any explanation; but many
perſons are ſo inſenſible to their uſe, as to ſup-
poſe that they impoveriſh the ground on which
they grow. This is by no means the caſe; they
thrive beſt in barren places, and love cold and
moiſture, and hence cover thoſe lands with verdure which would otherwiſe remain bare: ſo
far from injuring the plants, which are found
intermingled with them, they afford them
protection; their own roots penetrating to ſo
ſmall a depth into the ground, that they take
from it little nouriſhment; wherever a ſmall
quantity of graſs is found with moſſes, there

N   would

would be none without them; their leaves,
being of the kind called ever-green, continue
in vigour throughout the winter, and give
shelter to the roots of the grass which grows
beneath them. In spring the stems of the
moss, like all other evergreens, become bare,
and the ground is spread over with a fine verdure from the grasses which at that season
begin to vegetate; and if the land is drained
and manured it will be evident that the moss
has been no impediment to the growth of the
grass, even at the time of it's most luxuriant
foliage, as it will soon disappear after the improvement of the land, and the grass will
flourish even during the months of winter. A
yet more essential use is derived from various
species of moss, which grow upon the sides
and shallow parts of pools and marshes; in
process of time their roots occupy the space
which was before filled with water, and in
their half-decayed state are dug up, and used
for fuel, under the name of peat; of the importance of which no one can be duly sensible
who can enjoy plenty of coal. It is not, however, from moss alone that peat is derived; so
that it must not have more than a share of
praise among other vegetables, several of which,

7   even

even whole trees, form the compofition of peat
beds. Young plants are covered with mofs
in order to preferve them from froft, or burn-
ing heat. The gardener wraps his newly-
grafted trees with mofs, as from it's power of
retaining moifture a long time without pu-
trifying it preferves them from the injuries of
outward drought, and prevents the juices of
the graft from evaporating. Since the time
of Linneus it has been well eftablifhed, that
the mufci, or moffes, have diftinct fructifications, though botanifts are yet divided in
regard to their fituation; but as thefe plants
now have excited general attention, a few
years will give us, I hope, a revifal of the
works of Linneus, with the improved knowledge derived from modern inveftigation:
already an improvement in the clafs Cryptogámia has, I believe, been attempted and received; which encourages us to hope, we
may fee, at no very diftant period, that divifion of extraordinary plants no longer a reproach to the fcience. At prefent, the outer
habits, and fituation as to the growth of
the flowers or capfules, are chiefly made ufe
of to diftinguifh the génera of moffes. Thefe
plants refemble pines, firs, and other ever-

N 2   greens

greens of that tribe, in the form and difpofition of their leaves, and manner of growth
of their feed-bearing flowers, which are generally formed into a cone. Moft of the
moffes are perennial and evergreen; their
growth is remarkably flow; their anthers,
from their firft appearance to the time of the
difperfion of their powder, continue from
four to fix months. In fome of the fpecies
the leaves are fmall and undivided, and have
no vifible foot-ftalk, or mid-rib; in others,
as in hy'pnum proliferum, they refemble the
fronds of ferns. Their ftamen and feed-
bearing flowers are fuppofed to be placed
apart; fometimes on the fame, and fometimes on different plants. The calyx, termed
by Linneus the calyptre, covers the tops of
what he called the ftamens. From the prefence or abfence of this cover, which falls
before the opening of the fuppofed anthers,
Linneus, after Dillenius, has diftinguifhed the
genera. After the veil, or calyptre, is taken
off, there is found another cover to the anthers, which Linneus calls the operculum, or
lid. This is a beautiful microfcopic object;
and, with the other parts of the fructification
of moffes, fhould be firft ftudied by the affiftance

ance of plates, and afterwards inveſtigated by the agreeable amuſement of microſcopic ob-ſervations. Before the parts of fructification are protruded, they may be ſeen by the aſſiſtance of powerful magnifiers encloſed within thoſe ſmall buds, which terminate the leaves of moſſes, and have the appearance of being only a continuation of them. Hedwig diſcovered, that the leaves, or ſcales, compoſing theſe buds, differed materially from the leaves of the plant, and conſiders them as true involucres to the parts of fructification. He has alſo ob-ſerved, that in the capſule-bearing moſſes, which have their cones ſituated towards their extremities, the leaves adjoining the fruit-ſtalk are much more beautiful than thoſe of the ſtems. Sometimes the inner leaves become gradually ſmaller, and thoſe neareſt the fructi-fication ſo very minute as to make it impoſſible to take them away without a microſcope. Theſe involucres, like the calyxes of many other well-known plants, grow larger as the capſules advance towards maturity. Hedwig gives ſo minute and particular an account of both the ſtamen and ſeed-bearing flowers of the whole family of moſſes, that, if he has not been deceived in his reſearches, we may

N 3   expect

expect foon to fee a greater progrefs made in
the knowledge of this difficult tribe of plants,
than fome years ago it appeared probable
would ever be attained; but as thefe refearches
were made by the affiftance of the moft
powerful magnifiers, and with every advantage that could be procured, much information will not be gained from his plates of the
natural plant. From Mr. Curtis's defcriptions and figures the fpecies delineated by
him may be clearly underftood. He recom-
mends to the notice of young ftudents the
bry um undulátum, and curled bryum, as their
parts of fructification are large and diftinct.
Mr. Curtis does not pretend to decide the
queftion, whether the powder, from what is
called the capfule, is anther-duft, or feed.
Hedwig afferts, that thefe capfules are true
feed-veffels, and tells us, he fowed them, and
repeatedly procured from them a crop of young
plants, fimilar in all refpects to the parent plant.
Dillenius fowed thefe cones frequently, but
without fuccefs: it is probable that the fituation of the ftamens and piftils under one or
diftinct covers may have occafioned fuch different refults from the experiments of thefe
eminent botanifts. In the curled bryum,
the

the capfules or anthers are cylindrical, bent
inward, and if magnified they appear fomewhat ftriated. Their colour is firft green, then
livid brown, and laftly of a reddifh brown
colour. The bottom of the opérculum, or
lid, is convex and red; the top paler, very
flender, and rather blunt; the mouth of the
capfule is fringed, and the fringe bent inward; the ring is red, and the powder, which
iffues from the capfule, be it feed or anther-
duft, is green. Hedwig has obferved, that
this fringe of the capfule in dry weather
expands, and leaves the mouth of it open;
but on the leaft moifture, even of the breath,
it clofes again. He remarks, the ring of the
capfule of fome fpecies is elaftic; and, when
the feed is ripe, throws off the veil with more
or lefs force; and it is after this veil, or calyptre, is gone, that the fringe ferves to protect
the precious contents of the capfule. The
calyptre in bry'um undulátum is of a pale
brown colour, terminating in a long point,
firft upright, but afterwards, on the bending
of the capfule, it burfts at the bottom, and
remains ftraight, with it's bafe at fome little
diftance from the capfule *.

* A plate is given of the different parts of moffes for thofe
who have not the advantage of confulting Mr. Curtis's London
Flora.

The

The mechanifm of the fructification of the moffes, and that of the ferns, is truly admirable. Both feem intended for the formation, protection, and difperfion, of their feeds, or of fome fubftance equivalent to it; but, unlefs we credit the plates of Hedwig, we are equally ignorant of the manner in which this feed is produced in both tribes. In the magnified leaf of the bry'um undulátum the circumftance may be feen which has given it's fpecific name, the leaf being waved at the edge. This mofs produces it's fructification from November to February, and is commonly to be found either in woods or on heaths; it's leaves foon curl up, after the plant is gathered; feldom more than two peduncles arife from one ftem, generally only one; they are both longer than the ftem, upright, and of a reddifh colour.

Mr. Curtis has given a beautiful fpecimen of a mofs, which he has thought proper to place under the bry'um genus, although arranged as a mnium by Linneus. On the firft view it is diftinguifhable from the bry'um undulátum; it's bending peduncles, which have occafioned it to be called the fwan's-neck bry'um, are an obvious character in this fpecies; added to this, is the ftar-like appearance,

ance, which terminates thofe ftems from which the capfules do not proceed: thefe ftars are fuppofed, by fome authors, to be the female parts of fructification. Mr. Curtis, with very accurate inveftigation, was not able to difcover any thing in their ftructure, in the leaft fimilar to any of the parts of fructification in other plants. Hedwig afferts, that thefe ftar-like appearances are the involucres of the ftamen-bearing, or male flowers, and makes no doubt of the capfules containing the piftils, or female flowers. If the ftars and capfules are really diftinct parts of the fructification, it feems probable, from the fituation in which they grow, that the ftars contain the females, and the capfules the males; or fome of the genera of moffes may poffibly have flowers of all kinds, like thofe plants which compofe the clafs Polygámia. On this obfcure fubject I have thought it neceffary to give fome idea of the opinions of different botanifts, left, by detailing only the defcriptions of particular individuals, I might lead my readers to form too decided an opinion upon a point, which is not.yet fufficiently clear to juftify any thing further than conjecture,

The

The examination of two other kinds of mofs will give a pretty good idea of the parts which the young ftudent may expect to find in their different génera. The one firft to be confidered is the hy'pnum prolíferum. The hy pnum and bry'um families are feparated by Linneus from the fituation of the peduncle, which fupports what he terms the anthers, but which later writers have agreed to call the cap-fule. This in the bry'um grows out of the top of the ftem, and is furnifhed at it's bafe with a little naked tubercle, or bulb. In the hy'p-num the peduncle grows out of the fide of the ftalk; and the tubercle at it's bafe is co-vered with leaves. This elegant fpecies of hy'pnum derives it's fpecific name, *proliferous*, from the fingular ftructure of it's leaves, or fronds; one large fhoot proceeding from the middle of another repeatedly; and thefe fhoots extending themfelves along the ground, and taking root. Linneus found this beautiful plant in one of his journies through Sweden, growing in the thickeft woods, obfcured by perpetual fhade, and where no other plant could exift. This plant is not often found in a ftate of fructification, though by diligent fearch it may be fo. It's time of fructifying

is

is from December to February. The ſtructure
of the capſules will be found nearly the ſame
in all the moſſes. Mr. Curtis has, however,
diſcovered ſome peculiarities in thoſe of bry um
ſubulátum, or awled bry'um, and in poly'trichum ſubrotundum, or dwarf poly'trichum,
which are worthy of further attention. The
bry'um, after it has loſt it's calyptre and opérculum, protrudes from it's moúth a ſubſtance,
which by magnifiers is found to conſiſt of a
number of filaments, forming a thin ſpiral
tube, looſe and unconnected at the top: this
tube may be ſeen through the tranſparent
opérculum, forming in it's young ſtate a
ſmall ſpiral line. Mr, Curtis does not even
conjecture what may be the uſe of this extraordinary appendage; it may perhaps be the
receptacle of the ſeeds within the capſule,
which, on arriving at maturity, burſts open
the covers, and diſperſes it's contents. To
aſcertain this, there ſhould be ſowed repeatedly
a great number of theſe capſules, with and
without the tubes, and the tubes without
the capſules. There would, however, be great
nicety in the time that theſe capſules were
gathered: it is poſſible that, at the moment
of protruſion, the vegetating power may be
loſt;

loft; it fhould, therefore, not be too haftily
concluded that it did not refide in thefe filaments becaufe young plants are not obtained
from them; or if the capfules are fowed,
while their covers remain, and give no produce, it cannot be decided that they were
incapable of doing fo, as they might not be
in a ftate fufficiently mature.

The beautiful and curious ftructure of the
capfules of the poly trichum fubrotundum
are well worthy of the trouble of inveftigation, particularly as Mr. Curtis has found
their peculiar conftruction to be a conftant
character belonging to the genus, fo far as he
examined thofe fpecies which he could procure.
The capfules of moffes in general have only
one veil or calyptre; in this genus there are
two within the woolly calyptre of the poly'trichum, which has the appearance of a little
diftaff covered with flax. He found a membranous fhining fubftance, clofely connected
by it's top to the infide of the woolly one,
which is peculiar to this genus, but which
was fcarcely vifible, except by totally inverting
it; by doing fo, it is vifible to the naked eye.
This inner calyptre differs very little from
that of other moffes; at firft it wholly fur-

rounds

rounds the unripe capfules; as they increafe in fize, it fplits at the bottom, and finally becomes very fhort.

The beauty and curiofity of the ftructure of the capfules of moffes, with their whole elegant apparatus, may have detained me too long upon this fubject; but it is my wifh, by interefting my readers in the hiftory of their outer habits, to induce fome of the more inquiring among them to enter upon an accurate inveftigation of their parts and properties. If the account I have given of fome of the génera is in any degree found amufing, it is to Mr. Curtis I am indebted for the power of having made it fo. To thofe who can have accefs to his accurate and elegant plates, with his obfervations thereon, the clafs Cryptogámia muft be peculiarly interefting. But his London Flora being a work of too much expenfe to be of general utility, I am happy to have it my power to recommend to my readers the figures and obfervations on this difficult clafs, which may be found in Dr. Smith's Englifh botany. To his accurate defcriptions by the pen, and thofe of the pencil by Mr. Sowerby, we owe much information on the algæ tribe, which is now to

be

be explained. The plants comprifed under the defcription of algæ, or flags, fcarcely admit of a diftinction of root, ftem, or leaf; much lefs are their flowers fufficiently obvious to admit of a definition of their parts, though, by the fituation of their fuppofed flowers, or feeds, the génera are diftinguifhed, or fometimes by the refemblance of the whole plant to fome other fubftance with which we are familiar in the economy of nature. This tribe of plants is of great importance, as they frequently afford the firft foundation, from which other plants draw nourifhment. One fpecies of byffus, and feveral fpecies of lichen, fix upon the bareft rocks, and are fupported by what flender fupply the air and rains afford them. Dr. Smith, in his tour on the continent, in the years 1786 and 1787, found near Mount Vefuvius, on a torrent of lava, which iffued in 1771, the lichen pafchális, which covered it moft copioufly, and had the appearance of hoar froft, with no other plant near it. The lichen pafchális is peculiarly fitted for the beginning of vegetation on the hard furface of lava, from it's fhrubby figure, and flender roots; in the fame manner, the thread-form lichens infinuate their roots into crevices in

the

the bark of the oldeft trees, while the broad cruftaceous kinds. cover young bark, and the fmoother forts of ftones and rocks. The lichen pafchális being a perennial of very flow growth, many years elapfe before it's crumbling branches fall into the cavities of the lava, and there decaying form vegetable mould for the nourifhment of other plants. By attentive obfervation the progrefs, in which fuch vegetable mould is formed, may be feen on the fmooth and barren rocks upon the fea-fhore; and by a knowledge of the decaying plant we may know that, which will next fucceed. After the by ffus and feveral fpecies of lichen have crumbled into duft, firft appear other fpecies of lichen, which require a deeper foil for their fuftenance. When thefe perifh, and have again more thickly covered the rocks with mould, various kinds of the moffes appear; in their turn thefe alfo decay, when their places are fupplied by other plants, till a fufficiency of earth is accumulated to afford nourifhment to the largeft trees. It has been before obferved, that fome of the fpecies of lichen are ufed in dying; one of them, *lichen roccella*, called the orchel or argel, is brought from the Canary iflands, and forms a confiderable

derable article of traffic. They are a grateful food to goats, as well as to the rein-deer.

That beautiful vegetable called the cup-moſs is the líchen pyxidátus, or box-lichen. There is great difficulty in aſcertaining the ſpecies or varieties of the numerous plants of this genus. According to Hedwig's inveſti-gations the cup and ſaucer-like appearances, which are found on the various ſpecies of lichen, are to be eſteemed the ſeed-bearing flowers; and the notches, and warts with black tops, thoſe which contain the ſtamens. He aſſerts, that the fringes from the líchen ciliáris, fringed lichen, which take root, and the downy matter on the ſurface, have no-thing to do with the real parts of fructifica-tion. He gives very particular accounts of theſe parts, with plates of ſeveral génera of the algæ tribe; but the whole of theſe plants is at preſent ſo little underſtood, that it is not eaſy to give any accurate information con-cerning them. It is poſſible that too perti-nacious an inquiry after the mode of ſeminal reproduction in all the orders of the Crypto-gamia claſs may tend to retard rather than accelerate our knowledge on the ſubject. The plant called ſea-wrack is of the algæ

tribe,

tribe, and of the fúcus genus; it has it's
fpecific name of vesículous or bladdered, from
the bladders which cover it's furface. If the
leaves of this vegetable receive an injury or
fracture, while the plant is in a vigorous ftate,
abundance of young leaves are thrown out
from the injured part; even if a fmall aperture
be made in the middle of a leaf, a new one
arifes from either fide of it.

This fpecies of fúcus is frequently feen with
black hairy tufts, like horfe-hair, which are
commonly fuppofed to be a part of the plant;
but this is not the cafe; thefe tufts are diftinct
vegetables of the conférva genus, which attach
themfelves to the bladder fúcus, and appear to
belong to the plant itfelf. There are fome fpecies of fúcus which perhaps, on further inveftigation, may be found to partake more of
the animal than of the vegetable kingdom, in
the fame manner as the fea anemóne; which
was believed, till lately, to belong to the latter. The green fcum, which we fee on ftag-
nant water, and the green films on trees, are
but juft now beginning to be properly in-
quired into. In a courfe of years the whole
clafs Cryptogámia muft undergo a different
arrangement; and there is not any one of

O   the

the four orders, of which it confifts, requires it more than that which is now under confideration; neither can there be found, amongft the génera contained in it, a common chara&er ftrong enough to affemble fuch a variety of familics, which apparently differ in many ftriking circumftances: they all feem to poffefs peculiarities, which are well worthy of inveftigation; the beauty of the lichens attra&s our notice in winter on every tree, and bank, and wall, as they form a confpicuous part of that elegant arrangement, which is always found in an affemblage of the Cryptogámia families. That beautiful little plant, which is feen on heaths, and commonly called white mofs, is the rein-deer lichen; a knowledge of it's ufe to the ftarved inhabitants of the northern climates gives us an intereft in it, even beyond what neceffarily arifes from it's elegance of ftructure. There are many varieties of this plant, from which the true fpecies is diftinguifhable by it's very different appearance, although found in the fame places. The lichen fylváticus, wood lichen, which is only a variety of the rangiferínus, has uniformly it's branches of a reddifh brown colour, and it's ftalks fmaller, and fometimes

befet

befet with minute crifp leaves, and the whole
plant with age turns brown; neither of which
ever happens to the rein-deer lichen, it's colour always being white. What is commonly
called mofs on trees, is alfo a lichen. This
elegant tribe of plants well repays the trouble
of inveftigation; and, with the moffes, ferns,
and fungufes, furnifhes the botanift with a
complete winter garden.

The fourth and laft order of the clafs Cryptogámia contains the fungi, a tribe of vegetables, which, although they cannot vie with
the filices, mufci, or lichens, in beauty or
elegance, are not deftitute of either, and, from
the curious mechanifm of their ftructure,
cannot fail to intereft an inquiring botanift.
Mr. Curtis's, Mr. Bolton's, and Monf. Buil-
lard's plates will be great affiftants in the
ftudy of thefe vegetables; alfo Mr. Sowerby's
collection of fungi will be found highly ferviceable. The delicate botanift turns away
with difguft from the fmell and difagreeable
touch of fome of the fungi génera; but the
generality of them may be diffected by perfons of the greateft nicety without giving
offence. Within the laft twenty years our
knowledge has been greatly improved in regard

O 2   to

to the fructification of the fungi, as well as
that of the other three orders of the clafs
Cryptogamia, but yet remains fo imperfect,
that their generic characters continue to be
taken from their outer form. Hedwig's refearches tend to eftablifh for a fact, that the
fungi poffefs all thofe parts of fructification
which, in botanic language, conftitute a flower,
viz. ftamens and piftils. The ftamens he con-
ceives to be a collection of pellucid fucculent
veffels, with which innumerable oval globules
are connected, of a dilute brown colour.
Thefe fmall bodies he difcovered under what
is called the curtain, a part which is found in
fome fungufes, and not in others. This is a
thin membrane extending from the ftem to
the edge of the hat, which is torn as that expands, and foon difappears; but the part attached to the ftem often remains, and forms a
ring round it. The parts fuppofed by Hedwig to be the piftils, he found, in examining a
portion taken from one of the gills, which he
divided with fome difficulty into two plates,
the lower edge thickly fet with tender cylindrical fubftances; fome with globules at their
extremities, and fome without: the gill itfelf
appeared netted with larger and more diftinct
fpots,

ſpots, a little raiſed. In another fungus, a
ſpecies of agaric, after the curtain was torn,
and the hat pretty fully expanded, with the
gills turned yellow, he found the upper part
of the ſtem beginning to be tinged by a brown
powder, ſhed from the gills. On examination he did not ſcruple to pronounce this
brown powder to be the ſeeds, and that it
proceeded from the larger ſpots, that he had
before obſerved in the gills; the two folds of
which now readily ſeparated. He aſſerts,
that he has uniformly found in the génera of
agáricus and bolétus the globules, which he
believes to be ſtamens, either on their upper or
inner ſurface. In thoſe agárics, which have
neither curtain nor ring, theſe globules, with
their threads, are placed upon the ſtem.

Having given a ſketch of the modern diſ-
coveries in theſe obſcure vegetables, the outward habits and ſtructure of the fungus tribe
may be examined; and from the variety in
theſe circumſtances the ſtudent may endeavour to gain ſome knowledge of the characters
of the different génera. The reſearches of
Hedwig having been made with glaſſes of
highly magnifying powers, the parts which
he has diſcovered can never ſerve for the

O 3　　　　　diſtinction

diftinction of the génera; in which the character being obvious and clear conftitutes the
excellence of it. It is, however, very defirable, that fuch refearches fhould be made.
It is a decided fact, that fungufes continue
their fpecies by a powder, which is vifible
in the gills of many of them, and which is
generally allowed to be feed. Some fpecies
of the agáricus have fo fhort an exiftence,
that from the time of their appearance to
the time when they begin to decay, is not
more than five days. The manner in which
many of them decay, is by their gills diffolv-
ing into a very black liquor, like ink, that,
dropping, carries with it the feed; which
may be feen in the liquor, if greatly magnified. The ftructure of one of this genus fhould
be inveftigated, as it is the moft numerous of
the fungus tribe, and, if well underftood,
will bring the ftudent acquainted with the
bolétus, and other génera of this order. The
agarics are compofed of a pileus, or hat with
gills underneath, and with or without ftipes
or ftems, the pofition of the ftipes being
either central or lateral; from which arife
the three firft divifions of the genus; they
have alfo a root, more or lefs obvious; and

3   fome

fome of them, while in their unfolded ftate,
are wholly enclofed in a membranaceous, or
leathery cafe, called the volve. This cafe muft
not be confounded with that part fo termed
by Linneus. Mr. Bolton has fhown us the
juft diftinction betwixt the volve, and the
veil or curtain, the latter being what Linneus
has marked as the calyx, under the term
volve; which has occafioned a confufion in
thefe two parts, though in reality none can be
more evidently diftinct, applicable to different purpofes: the volve wrapping round
and protecting the whole plant in it's infant ftate; the veil apparently belonging to
the fuppofed parts of fructification only,
which Hedwig afferts he has found under
it. From the remains of the veil a ring is
formed: this part is not only uncertain in it's
time of duration, but even will appear in
fome years on the ftipe, and not in others;
confequently it cannot be ufed as a permanent character. The ftem of an agáricus is
either folid or hollow; the folid ftem differs
much in it's degree of folidity, fometimes
being as folid as the flefh of an apple, and
fometimes perfectly fpongy. Next to the
gills, the ftem of an agaric is the part leaft

liable

liable to vary. The gills are the part commonly known by that name, and with which every one is acquainted; they affume different colours in different fpecies, and vary much in their refpective lengths; each gill confifts of two membranes, and between thefe the feeds are formed; the gills are always attached to the hat, and fometimes to that only; fometimes they are not only fixed to the ftem, but extended along it downwards, like the wires of an umbrella. This has been called a *decurrent gill*. Mr. Curtis difcovered a peculiarity of ftructure in the gills of the agáricus ovatus, which he had not before obferved in any other fungus: the gills are connected together by numerous tranfverfe bars, or filaments, the ufe of which feems to be to keep them at an equal diftance from each other, and thus to admit the air to the fructifications, which are fituated on the flat furface of the folds, and to prevent their being deftroyed by preffure from their too great clofenefs. Thefe bars make it extremely difficult to feparate one of thefe folds entire: they are vifible only when greatly magnified. The fecondary fubdivifions of the agarics are founded upon the folidity or hollownefs

of

of their ftipes with the pofition of their gills, which, being the part wherein the fructifica- tions are contained, is of the greateft import- ance. They vary much in almoft every circumftance belonging to them, except in colour, which in all other plants is the moft variable of all their characters; the colour of the gills, on this account, is the mark, which has lately been adopted for the diftinction of the fpecies: their colour is fuppofed to be principally, if not wholly, caufed by that of the fructification or feeds, and is faid to have been found fufficient, with their ftructure, to afford permanent fpecific diftinctions. Thefe colours change, when the plant begins to decay; and of thofe agarics, which diffolve away in an ink-like liquor, the gills in their young ftate are white; fo that, to judge of their colour, the plant muft be gathered in it's firft ftate of expanfion, when they will be found to be gray. It is the colour of the flat fide of the gills which muft be attended to in the fyftem I am explaining to you, becaufe the colour at the edge in fome plants is dif- ferent through all the ftages of growth; and in others, it changes fooner than that of the fides, evidently from the difcharge of the

feeds,

feeds, when ripe. The hat of the agarics is leaft to be depended on; it's fhape is either conical, convex, flat, or hollowed; the top like a funnel. It is conftantly varying in the fame plant before expanfion, but not very changeable in the fame fpecies, when it is nearly, or fully expanded. The colour of the hat is extremely uncertain, therefore can only be attended to as a mark of varieties. The vifcous juice on the hat and ftipe, which is feen in many agarics, differs, according to their fituation, or to the ftate of the atmo-fphere, fo much, that the fame fpecies will fometimes be found glutinous, and at other times perfectly dry. Some of the agarics contain a milky juice, more or lefs acrid: this circumftance is not conftant, it having been found in the agáricus rubefcens, and the agáricus cæfareus, that plants equally vigorous, and in the fame fituation, will fome of them pour out milk in abundance on being wounded, while others will not exhibit any marks of it.

Upon the principles here explained, the late Dr. Withering has given to the world an arrangement of the fungufes, from which the génera may generally be inveftigated, It muft
be

be remarked, that an exception to the uni‐
formity in the colour of the gills takes place
in the agáricus aurantius, which fpecies ex‐
ifts under every kind of colour that can be
imagined. There is a variety of the agáricus
integer, entire agaric, which has it's hat of a
blood-red colour, and which appears from
Auguft to October. The colour of many of
the fungufes is beautiful; the moft fplendid
of all the agarics is the cæfareus, which in
England is a rare plant, but is common in
Italy, and brought to the markets for fale.

The plant we eat under the name of
mufhroom, is the agáricus campeftris, which
the gardeners propagate, either by fowing the
gills, or by planting fmall fibrous fhoots,
which are found about the bafe of the ftipe,
and which produce tubercles, in the manner
of potatoes. It may be difficult to affign a
reafon for the exclufive preference given by
the englifh to this fungus, as an article of
cookery. The caprice of mankind in their
choice or rejection of particular kinds of food
is not eafy to be accounted for. The agáricus
campeftris, however, feems to juftify the dif‐
tinction that has been given to it, as an efcu‐
lent vegetable, from the finenefs of it's flavour,

and

and tendernefs of texture. But although we make ufe of it at our tables, almoft exclufively, it has not the fame preeminence in other countries; and the inhabitants of Ruffia devour almoft every fpecies, even thofe which by other nations are efteemed poifonous. The noxious qualities of mufhrooms may be doubted of. Inftances of injury from the culinary ufe of the fungi tribe are certainly rare; and when they have occurred, it has remained doubtful, whether the poifon proceeded from the mufhroom, or from the veffel in which it was dreffed. But as mufhrooms make a part of our diet more palatable than nutritive, it can never be neceffary to eat them; and particularly if they are found hard it will be prudent to refrain from doing fo, as it is probable the poifonous effects recorded of them may fometimes have arifen from want of fufficient ftewing; for we have daily experience of the falutary ufe of fire to many of our vegetables, which in their frefh ftate would be fo far from affording wholefome food, that they could not be eaten without producing pernicious confequences. And the difufe of any particular fpecies of diet is of lefs confequence to highly civilized nations, whofe

whofe luxurious inhabitants have articles of
food procured for them from every quarter of
the world, and can thence form but faint
ideas of the neceffitous fituation under which
many of the inhabitants of the globe exift,
and in comparifon of whom our pooreft cot-
tagers may be confidered in a ftate of eafe.
In the rigorous and unfertile climates of
Sweden, Lapland, and Kamfchatka, that neceffity obliges the inhabitants to make ufe of
the inner bark of the pinus fylveftris (fcotch
fir) for food. In the fpring feafon they
choofe the faireft and talleft trees, and, ftripping off the outer bark, they colleƈt the foft
white fucculent interior bark, and dry it in
the fhade. When they have occafion to ufe
it, they firft roaft it at the fire, then grind it,
and after fteeping the flour in warm water to
take off the refinous tafte, they make it into
thin cakes, which are baked for ufe. The
poor inhabitants are fometimes conftrained to
live upon this food for a whole year, and are
faid to be fond of it; and it fhould be
nutritive, as Linneus afferts that it fattens
fwine. Nor ought we alone to eftimate the
vegetable tribes by the ufe to be derived from
them to the human fpecies. The fungufes,
                                    which

which are apt to be regarded in too infig-
nificant a light, afford fuftenance to a numerous fwarm of the animal creation, a variety
of infects. Although the pinus fylveftris is
unknown to more genial climes, as affording
an article of food, it has been applied by
mankind to more ufes than moft other trees.
The talleft and ftraighteft are taken for the
mafts of fhips; the timber is refinous, dur-
able, and applicable to many domeftic purpofes; fuch as making floors, wainfcots,
boxes, and all thofe things which are made
of deal; which is the name given to the
wood of this fir-tree, when fawn into planks.
From the trunk and branches of this, as well
as of moft others of the pinus tribe, tar and
pitch are obtained. Barras, Burgundy pitch,
and turpentine, are acquired by incifion. In
the highlands of Scotland, the refinous roots
are dug out of the ground, and divided into
fmall fplinters, which are burnt by the inhabitants to fupply the place of candles.
The moft important ufe, we have obferved,
is made of the inner bark by the Swedes,
Laplanders, and Kamfchatkans; of the fame
material, the fifhermen at Lockbroom in
Rofsfhire make their ropes. This fpecies of
fir

fir has acquired the name of fcotch, from being the only one which grows naturally in Scotland. It is found fcattered in many places amongft the Highland mountains; and large natural forefts of it are feen of many miles extent in various Lowland diftricts. From the cones of this fir a refinous oil is extracted, which is faid to poffefs virtues fimilar to thofe of the balfam of Peru. This tree lives to a great age; Linneus affirms four hundred years. The anther-duft in fpring has been carried away by the winds in fuch quantities, as to have alarmed the ignorant with the idea of a fhower of brimftone.

The laft genus of the Cryptogámia clafs to be confidered is mucor, or mould. It would fcarcely be fuppofed, that the mould found on bread, fruits, leaves, and various other fub-ftances in a decaying ftate, was a plant fubject to all the laws of the vegetable kingdom. That it is a plant of perfect form may be feen by the affiftance of a microfcope of common magnifying powers. It will be found growing in clufters; the ftems a quarter of an inch high, pellucid, hollow, and cylindrical; each fupporting a fingle globular head, which at firft is tranfparent, afterwards dark gray; thefe

heads

heads burft with elaftic force, and eject fmall round feeds, which are eafily difcoverable by the microfcope. It is the mucor mucedo which is here defcribed; but there are thirteen diftinct fpecies of mould, or mucor, which appear at different times of the year; one kind, called the golden, from it's brilliant yellow colour, covers the whole furface of plants, on which it grows, and ftains the fingers yellow, if touched. It is generally found upon the plants belonging to the bolétus family, and has the property of repelling moifture. It is faid to remain free from wet, though immerfed in water for a year. Great indeed are the wonders of nature in all her works, and in none more than in thofe of the vegetable kingdom!

EXPLANATION.

EXPLANATION OF PLATE IV. PART II.

FRUCTIFICATIONS OF MOSSES.

Fig. 1. A Plant of Bryum Undulátum of the natural fize.

Fig. 2. The Capfule much magnified with it's Calyptre.

Fig. 3. The Calyptre feparated from the Capfule.

Fig. 4. The fringed Mouth of the Capfule.

Fig. 5. . The Fringe, with the Ring taken off the Capfule.

Fig. 6. The Opérculum of the Capfule.

Fig. 7. A magnified Leaf of Bryum Undulátum.

Fig. 8. A Plant of Bryum Hórnum, Swan's Neck Bryum, to fhow the Rofe or Star which terminates fome of the Leaf-ftems, a.

Fig. 9. A Plant of Hypnum Prolíferum, to fhow the manner of it's-leaves growing out of each other, and of the Capfules being placed on the Stem, b.

Fig. 10. A Leaf greatly magnified, to fhow it's granulated appearance.

Fig. 11. The Capfule with it's Fringe. c, The Opérculum feparated from the Capfule.

Fig. 12. The Fringe with it's Ring, feparated from the Capfule.

Plate 4. *Part II. P. 220.*

Fig. 1.

Fig. 2.

Fig. 3.

Fig. 4.

Fig. 5.

Fig. 6.

Fig. 7.

Fig. 8.

a

Fig. 9.

b

Fig. 10.

c *Fig.* 11.

Fig. 12.

London, Published May 1.1797, by J. Johnson, St Pauls Church Yard.

LECTURE V.

On the Grasses.

HAVING proceeded regularly through the Classes, Genera, and Orders, with their different subdivisions, the young botanist will find some assistance necessary in the study of the graminiferous tribe of vegetables. This elegant assemblage of plants requires a peculiar mode of investigation; but that mode well understood, and the method of accurately dissecting them adopted,. it will not be found difficult to obtain a competent knowledge of their structure. The term Grass, as it is vulgarly used, conveys only a vague idea; and a common observer is surrounded in his walks by a variety of species, while he is not conscious of the precise existence of a single individual. It is only of late years that this useful and curious tribe of plants has been attended to; so that the knowledge of the most common and valuable vegetables of the creation is yet in it's infancy. They have been confounded under one common name in general,

and

and the few, which have been diftinguifhed
by a particular appellation, are far from being
univerfally known by it. Mr. Curtis, in this
part of the vegetable kingdom, as in every
other, has applied his refearches to the moft
ufeful purpofes. He has attracted the notice
of the rich by his more fplendid delineations
of a variety of graffes in his London Flora;
while he has diffufed through all ranks a
knowledge of thofe génera, which are every
where to be met with, by the low priced publication of his Practical Obfervations on Britifh
Graffes; a work from which a general knowledge of the outer habits of our moft common meadow graffes may eafily be attained.
This tribe forms one of the natural orders of
Linneus, and poffeffes a variety of common
characters, by which feveral forts of corn are
arranged with thofe génera, which are more
commonly known by the name of graffes. There
will be found a ftriking agreement in the parts
of fructification of all the graffes which may
be examined; but this is not more remarkable than the fimilarity of their general air,
their manner of growth, and their whole appearance. A fimplicity of ftructure characterizes the whole clafs; they have uniformly

formly a fimple, ftraight, unbranched, hollow ftem, ftrengthened with knots at certain intervals; this, which is commonly called the ftraw in corn, is termed by Linneus the Culm. At each knot there is always a fingle leaf, which ferves as a fheath to the ftem to fome diftance; when it fpreads out into a long narrow furface, of equal breadth all the way, till it approaches the end, where it draws off gradually to a point. The leaf is invariably entire in every fpecies, has neither veins nor branching veffels, being only marked longitudinally with lines parallel to the fides, and to a nerve or ridge, that runs the whole length of it. Another curious circumftance, almoft peculiar to this tribe of plants, and common to them all, is the feed not fplitting when it germinates, but continuing entire, till the young plant is fufficiently nourifhed by it's mealy fubftance to feek it's own food; at which time there remains of the parent feed only the dry hufk. Thefe plants are termed by Linneus one-cotylédoned, or one-lobed. In wheat this may be well feen; and if the feed is preffed betwixt the fingers, when the plume has rifen an inch or two above the ground, it will be plainly

perceptible

perceptible that the fkin only remains. The common meadow fox-tail will fhow the peculiarities which may be found in the whole order of the graffes; and it is better to ftudy their characters in the natural plant than in plates; although Mr. Curtis's London Flora will afford much amufement and information upon the fubject. Upon examining the leaves and fheaths by a microfcope, many of them will be found furnifhed with briftles, which give them the appearance of a faw; from this circumftance, or the contrary, the fpecies are frequently diftinguifhed one from the other. The parts of fructification, from their want of fplendour, commonly pafs unnoticed, although their beauty and ftructure are fuch as muft excite our higheft admiration, when known. The natural character of the flowers of graffes is their having a glume, or hufk, which is the term given to their calyx by Linneus. This glume is compofed of one, two, or three valves, generally only two; the larger valve hollow, and the fmaller one flat. Thefe valves are a kind of fcales, with their edges commonly tranfparent, and moft frequently terminated by a pointed thread, termed by Linneus arifta, or awn. The

awn

awn is particularly ftrong in the hordeum genus, of which barley is a fpecies; but may be found in a lefs degree in various other génera, though not conftant through every fpecies; whence it's prefence or abfence is ufed by Linneus as a fpecific diftinction. The corol of graffes is alfo termed a glume, and in reality is only a dry fkinny hufk, confifting of two valves. The calyx and corol fhould be compared with a magnified drawing, and the natural parts looked at through a microfcope; their conftruction will then be underftood. The divifions of the outer glume, or calyx, ought always to be attended to, as it is often made ufe of by Linneus as a mark of the génera. Betwixt the glumes, or corol and calyx of the graffes, the young botanift may find himfelf perplexed; but it muft be remembered that thefe parts of fructification are not, in general, diftinctly defined at prefent; therefore they muft be underftood according as they have been diftinguifhed by Linneus. The inner glumes of the graffes are to be efteemed the corol, the outer the calyx. The flowers of this tribe have alfo univerfally a vifible nectary, confifting fometimes of two very fmall oblong leaves, placed at the

P 4   bafe

bafe of the germe, and fometimes different
kinds of fcales in the fame fituation, which
are diftinctly fhown in Mr. Curtis's plates
of the hólcus móllis, creeping foft grafs,
mélica uniflóra, fingle flowered melic grafs,
and mélica cærúlea, blue melic grafs, and
are not difficult to be feen in the natural
flowers. Though very minute, the leaves, of
which the nectaries are compofed, may be
feen at the bafe of the germe of the flowers
of wall-barley. Thefe leaves nearly refemble
the corol, but are lefs, and tranfparent; they
are named nectaries by Linneus; but as they
furnifh no generic diftinction, they are not
noted in the characters of all the génera. The
number of ftamens, that will generally be
found in thefe flowers, is three, with two
piftils, within the fame cover. But there are
exceptions to this rule, which fhall be explained prefently. The ftamens have three
hair-like filaments with oblong anthers of
two cells. The ftyles of the piftils are downy,
bent back, with their ftigmas beautifully feathered, in fome fpecies large and branching,
which, with the anthers waving on their long
filaments, form a moft elegant appearance;
but their parts are fo delicate and minute,
that

that they are feen to greater advantage if viewed through a microfcope. The clofe fpiked graffes do not fhow the parts of fructification fo well as thofe with loofer fpikes, or the panicled kind. In feather-grafs, ftipa pennata, they are very well feen, if examined in a proper ftate; but it is even more neceffary to inveftigate thefe flowers, before their anthers have difcharged their duft, than thofe of the other claffes; for as foon as the cafes containing it are burft, the whole plant affumes a withered afpect, and all parts, except the feed, fall to decay. Thefe flowers have no feed-veffel, and only a fingle feed; which is enclofed by either the calyx or corol, from which, when ripe, it is emitted in various ways. The twifting of the long awn of feather-grafs, in order to extricate itfelf from it's receptacle, which in this tribe is the ftem lengthened out to ferve that purpofe, gives it a very peculiar appearance. This will alfo happen if a bunch of the feeds be gathered, and bound tightly together; they will twine themfelves into all kind of directions, till they get loofe from the bondage which has been impofed upon them, and thus commit themfelves to the earth, where they vegetate and produce a

new

new progeny. The parts of fructification may
be well'feen in the flowers of the bríza máxima. The beautiful drooping fpikes of this
fpecies are peculiarly elegant from their tre-
mulous motion, caufed by their flender
peduncles, whence the genus derives it's
common name of quaking grafs. Although the characters here given of the
parts of fructification are all found nearly
conftant in thofe génera, which are placed
in the clafs Triándria, there are others
which fail in the claffic character of the
number of ftamens, and are thence placed by
Linneus in different claffes; which feparation
of plants, manifeftly of the fame natural order, is the more extraordinary, as, in fome
cafes, he has not thought it neceffary ftrictly
to adhere to the obfervance of the claffic character, when it has fo directly militated
againft an obvious fimilarity in every other
part of the fructification, as in hólcus lanâtus, but has made the difference the foundation of a fpecific character. The hólcus
lanâtus, meadow foft grafs, having fome of
it's flowers deficient in the proper number of
ftamens and piftils, which would rank it in
the clafs and order Triándria Digynia; Linneus

neus has torn it from all it's natural connec-
tions, and placed it amongft a tribe of plants,
in the clafs Polygámia, to which it has no
affinity. His moft flagrant faults, however, of
which this muft be efteemed one, admit of this
excufe, namely, the greatnefs of the work,
with which he has enlightend the botanical
world. We ought to be lefs furprifed, that
we find in it a few imperfections, than that
there are not more. This regarding the
hólcus may probably have efcaped, by fome
accident, his correction, as it is not uncommon to find the fame imperfection in the
flowers tríticum and hórdeum, wheat and
barley, and fome other graffes, which cannot
be confidered as conftant, but may arife from
a variety of caufes: and, as the character of
the claffes is purely arbitrary, it may admit of
a doubt, whether in all cafes it would not
have been better to have obferved it uniformly, than ever to have deviated from it.
So, for inftance, the genus anthoxánthum, which
in every particular agrees with the character
of the grafs tribe, except that of it's number
of ftamens, which are only two, and that
without variation. From this circumftance
Linneus has placed it in the clafs Diándria,
two-

two-ftamens. Had he done otherwife, a young botanift muft have found himfelf much perplexed; the claffic character being the firft that he would refer to, he could never find the anthoxanthum in a clafs, the effential character of which was three-ftamens, though, from it's general appearance, he could not expect to find it feparated from the reft of the graffes. There are fome peculiarities in the fructification of anthoxánthum odoratum which are worth attending to: a fpecimen of it fhould be diffected, and compared with a magnified drawing of it's different parts. It agrees with many other graffes in it's fmall fpikes, containing only one flower, but differs from the whole of the tribe in the following particulars: one of the valves of the glume, or calyx, is fmall and membraneous, the other large, and wrapping up, as it were, the whole of the fructification. Thefe glumes have been ob-ferved not to open and expand themfelves, as in the avéna genus, and other graffes, but the ftamens and piftils have the appearance of pufhing themfelves out of the glumes, which remain clofed; the glumes of the corol are not like thofe of other graffes, but are re-markably hairy, each having an awn, the

7   longeft

longeſt of which ſprings from the baſe of the
glume, and is at firſt ſtraight; but as the ſeed
becomes ripe, the top of it is generally bent
horizontally inward; the other awn ariſes
from near the top of the oppoſite glume or
valve. The nectaries alſo differ as much
from their common ſtructure, in this order
of plants, as the other parts of fructification;
they are compoſed of two little oval ſhining
valves, one of which is ſmaller than the
other: theſe cloſely embrace the germe, and
are difficult to be ſeen, unleſs they are ob-
ſerved at the moment of the anther's protruding from between them, at which time
they are very diſtinct: as ſoon as the anthers
are excluded, they again cloſe on the germe,
and form a coat to the ſeed, which remains
with it. The anthoxanthum is the graſs,
which gives the fragrant ſcent to hay; and
if the leaves are gathered, and folded up in
paper, they will retain their agreeable ſcent
for a long time: hence the ſpecific name
given to it by Linneus, of odorátum. It has
been ſaid to be the only engliſh graſs that
has fragrance; and this may be true reſpecting the leaves. But Mr. Swayne, in his account of paſture graſſes, informs us, that the
flowers

flowers of the annual poa have a sweet smell like those of the reséda odoráta, mignonette; and that the scent remains in the flowers when dried. The anthoxánthum is said to have two modes by which it is propagated; first, the common way, by seeds; and secondly, by bulbs formed upon it's stems, which fall off, when mature, and strike root into the ground. This circumstance, is said also to take place in many of the alpine grasses, by which means their species are preserved, which would otherwise be annihilated, so perpetually are their seeds devoured by small birds.

The seeds with which canary birds are fed are from a species of phálaris, deriving it's specific name, canariénsis, from the place of it's native growth, the Canary islands. The ribbon-grass is also a variety of another species of phálaris, the arundinácea, or reed phálaris, and makes an elegant appearance amongst the gayer colours of a flower-garden. The genus avéna, of which the common oat is a species, is obviously marked by a twisted and jointed awn, which issues from the back of the corol. The seeds of avéna fatua, fool's oat, or, as it is commonly called, wild oat,

exhibit

exhibit an amufing fpectacle. If placed on a table, after having been moistened in water, they twift themfelves about with fo much appearance of life, that the plant has been called the animated oat. There is alfo a curious circumftance belonging to the feed of barley: it's awn being furnifhed with ftiff briftles, which all turn towards the point, like the teeth of a faw, as this long awn lies upon the ground, it extends itfelf in the moift air of the night, and pufhes forward the barley-corn, to which it adheres: in the day it fhortens, as it dries; and as thefe points prevent it from receding, it draws up it's- pointed end, and thus, creeping like a worm, will travel many feet from the parent plant. The ingenious Mr. Edgworth con-ftructed a wooden automaton upon the prin-ciples of a barley-corn, which fucceeded fo well that it walked acrofs the room, in which it was kept, in the fpace of a month or two. Wheat, triticum hybérnum, the moft nu-tritive of the various grains which are applied to the ufe of food, is found in moft parts of Europe and Afia. Where the climate is too hot for it's cultivation, as in the torrid zone, it's place is well fupplied by what is commonly called

called India, or Turkey wheat, which is a
fpecies of zéa; a genus placed by Linneus in
the clafs Monœcia, one-houfe. Although
rice is ranked among the graffes in the natural orders of Linneus, he has feparated it
from them in his artificial fyftem, in confequence of it's being found deficient in the
effential character of his claffical arrangement
of thofe génera to which it bears fo near an
affinity. He has placed it in the clafs Monœcia. Rice is a fpecies of the genus ory'za.
In moft eaftern countries this grain is the
chief fupport of the inhabitants; and, fo far
as it is ufed for food, is wholefome and nutritive. But as we too often convert what, if
properly ufed, would be a bleffing into a curfe;
they make from it a fpirituous liquor, called
by the englifh arrack; which, like all other
fpirituous liquors, may be efteemed a flow
poifon. Moft of the plants belonging to the
natural order of graffes afford plentiful and nutritive food, not only to mankind, but to
beafts, birds, and infects, and have the remarkable property of not being deftroyed, though
continually trampled upon; indeed, they are
conftantly renewed by feeds; as their flowers,
the fame as in other plants, are never eaten by
cattle,

cattle, which, if left at liberty in the pasture, uniformly reject the straw on which the flower grows, devouring only the herb of the plant, so that the seeds which escape the small birds, ripen, fall to the ground, and renew their species. Those grasses which are more liable to have their seeds destroyed, or which, from the coldness of the climate they inhabit, cannot bring them to perfection, become viviparous, and perpetuate their species by a bulbous progeny. The similarity of calyx, corol, and nectary, in the grass génera, and the minuteness of their dimensions, will frequently prevent their being accurately distinguished from each other, till the student is become familiar with the appearance of all these parts; and he will then find them not more difficult of investigation than the fructification of many other plants.

EXPLANA-

EXPLANATION OF PLATE V. PART II.

FRUCTIFICATIONS OF GRASSES.

Fig. 1. A Spike of Alopecúrus Praténfis, Meadow Fox-tail.

Fig. 2. A Floret magnified. *a*, The Glume of the Calyx, with it's long Awn fixed to the Bafe. *c*, The Stamens. *d*, The Stigma.

Fig. 3. A Floret of the natural fize feparated from the Spike.

Fig. 4. The Stigma and Seed.

Fig. 5. The Germe and Styles of Póa triviális. *e, e*, The Nectary Glands.

Fig. 6. The Seed with a woolly fubftance at it's bafe.

Fig. 7. Part of a Spike of Anthoxánthum.

Fig. 8. The Stamens, Styles, and Seed, with the adhefive Nectary Glumes.

Fig. 9. The Nectary Glumes at the moment of protruding the Anthers.

Fig. 10. A Floret of Avéna Fatua, Animated Oat.

Plate 5. *Part II. P. 226.*

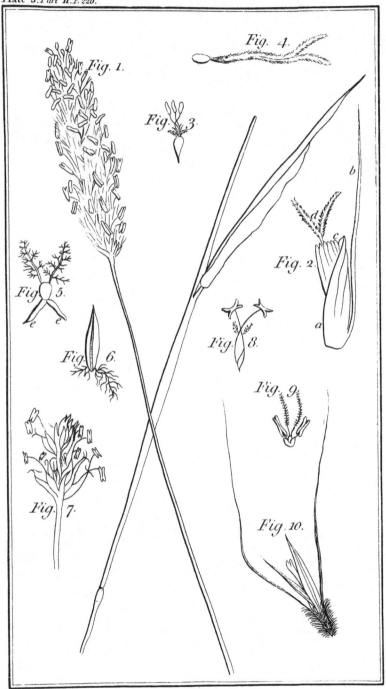

Fig. 1.

Fig. 4.

Fig. 3.

Fig. 2.

b

d

c

a

Fig. 5.

e *e*

Fig. 6.

Fig. 8.

Fig. 9.

Fig. 7.

Fig. 10.

London, Published May 1. 1797, by J. Johnson, St. Pauls Church Yard.

LECTURE VI.

Specific Diftinctions, and Double Flowers.

THE part which yet remains to be confidered of the Linnean fyftem is the fpecific
diftinctions, or thofe characters by which
every individual is diftinguifhed from others of
the fame genus. In this part of botany we are
even more obliged to Linneus for the order,
that he has introduced, than in any other.
He was the firft who began to form effential
fpecific characters. Before his time there
were no fpecific diftinctions worthy of notice; from which deficiency arofe great confufion. Now the knowledge of the fpecies
confifts in fome effential mark or character,
by which it alone may be diftinguifhed from
all other fpecies of the fame genus. Thefe
diftinguifhing characters are noted by Linneus after every individual of a genus; and
this is called the fpecific defcription. To each
fpecies he has given a name appropriated to
itfelf, which he has termed the Trivial Name.

Sometimes

Sometimes this name expreſſes ſome quality
of the plant, to which it belongs, but as frequently is arbitrary; and perhaps it would
be better that it was always ſo, as the names
by which we diſtinguiſh the individuals of a
family. It may require ſome trouble at firſt
to acquire the uſe of arbitrary names, but
the advantage of them when acquired is
every day demonſtrated. Of this we cannot
doubt, if we attend to the confuſion occa-
ſioned in common converſation, by perſons
who will not uſe the proper name of whatever they attempt to deſcribe: they introduce
all kind of circumſtances to make themſelves
underſtood, and at the end of their endeavours leave the perſon, whom they would
inform, in deſpair of ever acquiring any knowledge from their deſcriptions. Could the
diſtinguiſhing mark of each plant be ex-
preſſed by one word, and that word be uſed
as the name for the individual, or what is
called the trivial name, it would greatly fa-
cilitate the knowledge of plants; but this we
cannot at preſent hope, though probably we
ſhall ſee great improvement take place in this
part of the Linnean ſyſtem of botany, as well
as in ſome others.

It

It is defirable that all young ftudents in
botany fhould make a point of ufing the
terms and language of the fcience; and herein
will be found the fuperior excellence of the
Lichfield tranflation, that, in acquiring the
language of that work, we become able to
underftand any defcriptions of plants which
may occur to us in latin; whereas, when
there is an attempt made to form the terms
more after the englifh language, they cannot
be made ufe of except in converfation with
an englifh botanift: the fame objections occur againft forming either the generic or trivial
names in our own tongue in preference to a
literal tranflation of thofe given by Linneus.
One or two inftances will fhow the inconvenience of fuch a practice. Out of fix fpecies
of plantágo defcribed in the Botanical Arrangement of Britifh Plants, there are only
two which have their trivial names tranflated; fo that a ftudent, who formed his
language from that work, would find it almoft equally difficult to underftand a Linnean botanift, when he fpoke of plantago
media (middle), or plantágo lancéolata (lanced), one being termed hoary, and the other
rib-wort, as if he was ignorant of the fcience.

Q 3   Alfo

Alſo rúmex pulcher, or beautiful, has the trivial name fiddle given to it; and pulmonária officinális, officinal, is called broadleaved. Many more ſuch falſe names might be enumerated, which are equally awkward and injurious to the ſcience, and what every true botaniſt ought to avoid. I warn all my young readers ſtrongly from the uſe of ſuch terms, as they may hear them not unfrequently defended, as being more eaſy to acquire: but ſuch defenders are too idle to think much on the ſubject, and of courſe are little aware of the narrow extent to which their botanical knowledge can carry them, if founded only on the language of their own country, and of the plants contained in it.

But to return to the circumſtances from which Linneus has taken his ſpecific deſcriptions: he lays it down as a fundamental rule, that they are to be formed from ſuch parts of plants as are not ſubject to variation; great inconvenience having ariſen from the want of obſervance of this rule among former botaniſts; every variety being ranked by them as a diſtinct ſpecies. Colour is decidedly one of the leaſt permanent characters to be found in plants, conſequently not to be admitted into the

7   ſpecific

fpecific character. However, it muft be ac-
knowledged, that in contradiction to more
than one of his own rules, Linneus has made
ufe of colour, and other variable properties in
plants, to diftinguifh them individually one
from another. Linneus efteemed the root of
plants a true fpecific mark; but, from the difficulty of obtaining a fight of it, has never
made ufe of that part as fuch, if any other,
equally permanent and more obvious, could
be found. The trunk and ftalks of vegetables, in many inftances, afford fuch effential
differences, that they ferve to afcertain the
fpecies beyond a doubt. In the genus hypéricum, three of the fpecies are accurately
diftinguifhed by their ftems being round,
two-edged, and fquare. The different kinds of
inflorefcence and fulcra furnifh alfo permanent
marks. Linneus too has made ufe of parts
of the fructification for the purpofe of difcri-
minating the fpecies, which is done with good
effect in many inftances, though certainly in
a few cafes, in contradiction to the principle,
on which the claffes are founded, if confidered with ftrictnefs, as in fome of the
graffes; but where the characteriftic mark
of either clafs or order is not interfered

with, the parts of fructification form obvious
and agreeable marks of specific distinction,
as in some of the hypéricums, the species
are distinguished by their number of styles;
and in gentiána, the form and division of the
corols afford an obvious and permanent difference, which cannot be mistaken by the
most superficial observer.

But before the young student can hope to
arrive at a ready discrimination of plants, he
must study leaves under all their various
forms. It is from leaves that the most elegant and natural specific distinctions are
taken. Nature delights in variety in none
of her works more than in that of leaves.
The different sorts are exceedingly numerous,
and ought to be attentively studied by every
pupil in botany. In the present part of the
subject they are to be considered only as
marks of distinction, by which the individuals
of a genus are known from each other.
Their use and formation belong to another
part of the study. They must be taken methodically, and they will not then be found
difficult to understand, with the assistance of
the plates and botanical terms and definitions
given at the beginning of the System of
Vegetables.

Vegetables. The *form* of leaves is firſt to be conſidered, by which muſt be underſtood their external ſtructure. Reſpecting their form, they are divided into ſimple and compound leaves. Simple leaves are thoſe which have only a ſingle leaf on a petiole, or foot-ſtalk. Theſe ſimple leaves may differ in reſpect to many circumſtances, but they are ſtill ſimple, if the diviſions, however deep, do not reach to the mid-rib. There are ſixty-two ways in which a ſimple leaf may be diverſified, all of which muſt be ſtudied with the plates, and the terms of explanation annexed to them. The genius of Linneus is more conſpicuous in this part of his ſubject than even in any other. He has formed a language, which, in the moſt conciſe expreſſive manner poſſible, depictures ſuch a variety of forms of leaves, fruits, flowers, ſtems, and ſeeds, as no other was ever before made to deſcribe. The intro-duction of theſe excellent terms to engliſh botaniſts we owe to the Lichfield tranſlators of Linneus's works. To the Syſtem of Ve-getables are prefixed a preface and advertiſe-ment, which ſhould be read by all young botaniſts. Attention and habit will render the amazing variety of form in the ſimple

leaves

leaves familiar. The language of Linneus, as applied to the species of plants, must be studied, and may be understood without much difficulty. He has taken words expressive of well-known figures, as the words oblong and egg, which, simply used, signify that the leaf or seed is one of those forms; by compounding those words a form between both is expressed; if it partake most of the oblong, that word precedes the egg, and contrariwise; so that the two words, oblong and egg, are made to represent forms of four kinds very nearly allied. Thus has Linneus compounded all the different forms under which leaves can appear; and by having done so has been able, in a few words, to present before our eyes the essential specific characters of a variety of plants; which by other authors are described with so little precision, and so diffusely, that we are bewildered by the innumerable distinctions, to which we have to attend.

In order to attain a precise idea of these forms the student must begin by comparing the plates. The leaves of daisie (béllis) are oblong, those of beech (fágus silvatica), and pepper-mint (méntha piperita), egg-form, of

violet

violet heart-form, rofemary (rofmarinus officinális) and crócus, linear; or every where
of an equal breadth. When he has well
ftudied the fimple forms he muft then endeavour to underftand thofe which are compounded from them; and, by drawing,
compound the forms himfelf, till they become
familiar to him. Pulmonária officinalis, commonly called Jerufalem cowflip, has it's radical,
or root leaves, of the form betwixt egg and
heart; in expreffing which, and the reft of
the compound forms, the Lichfield tranflators have moft happily imitated the concifenefs of their author; and in their language
you will find the terms, egg-hearted, heart-
lanced, ufed inftead of between egg and
heart-fhape, heart and lance-fhape, and fo of
them all. The term arrowed is ufed for
arrow-fhape; lyred for lyre-fhape; twoed, or
threed, for growing two together, or three
together: indeed, inftances occur fo frequently of the agreeable concifenefs, with
which the language of the tranflated Syftem
of Vegetables is formed, that it would be difficult to enumerate them all: it is a work of
the higheft value to an englifh botanift. An
outline of the forms which may be found in
leaves,

leaves, both in their fimple and compound
characters, being underftood, thofe circumftances which conftitute a compound leaf
fhould be confidered. It has been fhown, in
treating of fimple leaves, that they continue
to be fo denominated, be their divifions ever
fo deep, provided thofe divifions do not extend to the mid-rib; but when that takes
place, the leaf becomes compound; fo that
it is in fact a fmall branch compofed of a
number of individual leaves, which feparate
leaves are frequently furnifhed with each a
petiole, uniting them to the common petiole,
or foot-ftalk; which, running through the
whole, is called the mid-rib. In fome inftances it may not to a young botanift be
very eafy to diftinguifh a compound leaf from
a branch; but there are two rules, by which
they may always be known afunder: ift,
buds are never found at the bafe of the lobes,
or divifions of a compound leaf, but are
formed in the angle made by the whole with
the ftem, from which it iffues; 2dly, the
branches of woody plants continue, after the
leaves are fallen: this never happens with a
compound leaf; for, however nearly the
common foot-ftalk, from which it is formed,
may refemble the other in appearance, it
always

always falls off, either with or after the leaves it fupports. The leaves of robinia, rofe acacia, afford a good example of the compound character, and alfo of the two rules that have juft now been mentioned. There are three kinds of compound leaves, the compounded, decompounded, and fuper-decompounded. The firft has been explained; and, although there be but two divifions from the fame common petiole, it is a compound leaf. The terms decompounded, and fuper-decompounded, are applied to different modifications of the compound leaf; and again thefe modifications admit of fuch a variety of others, which are diftinguifhed each by an appropriate term, that nothing but practice, and the method recommended in regard to the ftudy of fimple leaves, can bring the pupil acquainted with them. The feathered, footed, winged, paired, arc all different forms of the compound leaf; fo is the fingered, of which an example may be feen in the horfe-chefnut, æfculus hippocáftanum, and lupine (lupínus); as thefe various modes frequently enter into, if not entirely form, the fpecific character of plants, it is neceffary they fhould be well underftood. But, before the compound leaves are attempted,

it

it will be well to become perfectly acquainted with the different forms which exift in the fimple leaves; as the form of the fingle leaves, of which the compound leaf confifts, is a circumftance generally noted. The Syftem of Vegetables, methodically ftudied, will carry the ftudent through this difficult part of botany; or, if fometimes he may find himfelf perplexed, an explanation of the fame terms in other books will be of fervice to him, as he will probably find different words ufed, which may elucidate the point on which he may be in doubt. There are fome other circumftances relative to leaves, which it is equally effential to underftand as thofe which have juft now been treated of; thefe are, the determination, or difpofition of leaves, which comprehend four particulars alike belonging to the fimple and compound kind, the *place, fituation, direction,* and *infertion.* By the place, we are to underftand the particular part of the plant to which the leaf is attached. Situation regards the refpective pofition of leaves one to the other: fo leaves are called alternate, when they come out fingly, and are ranged gradually on both fides of the ftem, as in ivy toad-flax (antirrhinum cymbalária); or

oppofite,

oppofite, when they come out in pairs, as in myrtle (myrtus), and many other plants. Thefe two circumftances of leaves being alternate, or oppofite, furnifh conftant and invariable characters, which are generally found in plants of the fame genus, or even of the fame natural order. Direction contains the different ways in which a leaf bends from it's ftem; the various modes of it's doing fo are arranged under the general term direction, and muft be ftudied to be underftood. Infertion comprifes the diverfity of manner by which leaves may be attached to their parent plants, each of which has an appropriate term, briefly and expreffively explained in the botanic terms and definitions at the beginning of the Syftem of Vegetables, with plates at the end of each volume to illuftrate them.

I have now only to fpeak of fuch flowers as are commonly called double. To enter far into an account of them belongs rather to the natural hiftory of plants, than to that part of the fcience which ought to engage the attention of a pupil in the beginning of his ftudies. It will be fufficient to acquaint him with the unnatural varieties under which flowers appear, that he may not be mifled, by the monftrous

forms

forms they frequently affume, to look for a
genus where there is only a fportive variety.
Double flowers are the pride of a florift, as
they manifeft the art of culture; many of
them being formed by over luxuriancy of
nourifhment. Gardeners imagine, that by
placing a double ftock-flower near a fingle
one, they can thereby procure fuch feed as
will again produce double flowers: but that
this is a vulgar error, a very flight knowledge of botany may convince us; for, when
a flower is completely double, it is deprived
of it's ftamens, which commonly expand
into petals; by which transformation the
flower no longer poffeffes the anther-duft, or
effential part to the fertilization of feeds.
There are various ways in which vegetable
monfters are formed, moft of which generally
exclude all, or part of the ftamens. The
unchangeable parts of double flowers are the
calyx, and the lower row of petals, by which
the genus may be often difcovered. Some
flowers are only half-double; in which cafe
the ftamens and piftils often remain perfect,
and hence produce fruit. This happens in
the double peach, the fertility of which is
fometimes brought as an objection to the

Linnean

Linnean fyftem. There is one kind of the double, or multiplied flowers, which is termed proliferous; of this fort is the hofe in hofe polyanthos, and béllis prolífera, hen and chicken daify: this is one of the moft curious of vegetable monfters, as well as the moft beautiful. Plantágo rofea, or rofe plantain, is wonderfully difguifed by it's bracts becoming enlarged, and being converted into leaves. Many flowers become double by the multiplication of their nectaries, and in fo many various ways, that it would engage too much time to enumerate them. In the Provence rofe the petals are fo profufely multiplied as entirely to exclude the ftamens. In fome other rofes may be found ftamens, although the flower has a luxuriancy of petals, as in damafk rofe. The many-petalled flowers are the moft fubject to multiplication. The one-petalled rarely go beyond a double corol, which is very often feen in them. The compound flowers alfo are liable to become double; and their beauty is often improved by it; as daify, béllis, fneezewort, achilléa, and chryfánthemum sílphium; but, if we except a few inftances, I think fingle flowers are much to be preferred to double ones.

R   Befide

Befide the varieties occafioned by multiplication, there are others arifing from many
accidental caufes; but the moft general caufe
may be efteemed culture: it is from the gardener's art that we receive fo many delicious
fruits and vegetables for our tables; culture
too is the teft, whether a plant be a true
fpecies, or a variety. By a change of foil we
can produce the moft valuable varieties; or
oblige them to return to their original form,
by refufing them our nourifhing care. The
ingenuity and induftry of man is not feen
in any thing more confpicuoufly than in his
culture of corn, which, without the fcience of
agriculture, would be of fmall value; with it,
we muft efteem it the firft bleffing of life
Botanifts are careful to diftinguifh between
varieties obtained from feed, and the genuine
fpecies, from which they deviate. Such plants
will not be found noted in the Syftem of Vegetables, which contains only the génera, and
the permanent fpecies. In the Species Plantarum the varieties are diftinguifhed by a capital B being placed immediately before the
defcriptions of them. What has been explained refpecting the changes which take place
in the fructification of plants, is equally applicable

plicable to leaves, and to every other part of them; by which they are frequently ſo metamorphoſed, that it requires no ſmall degree of botanical knowledge to aſcertain the real plant. Many of theſe appearances may be effected by art, and have been ſo by the curious, in order to diſcover the true cauſe of ſuch deformities, or of diſeaſes, which are found deſtructive of vegetation.

THE END.

Printed by T. Benſley, Bolt Court, Fleet Street, London.

FLORIST'S MANUAL;

OR,

HINTS FOR THE CONSTRUCTION OF

A

GAY FLOWER GARDEN.

WITH

OBSERVATIONS ON

THE BEST METHODS OF PREVENTING THE

DEPREDATIONS OF INSECTS.

BY

THE AUTHORESS OF

BOTANICAL DIALOGUES, AND

SKETCHES OF THE PHYSIOLOGY OF VEGETABLE LIFE.

———————

Illustrated by Two engraved Plans.

———————

LONDON:

PRINTED FOR HENRY COLBURN, CONDUIT STREET.

———

1816.

W. Flint, Printer, Old Bailey, London.

DEDICATION.

THE FOLLOWING PAGES ARE INSCRIBED,

BY THE AUTHORESS,

TO HER HIGHLY ESTEEMED FRIEND,

LADY BROUGHTON,

AS A TRIBUTE TO THE TASTE AND INGENUITY WHICH SHE

HAS DISPLAYED

IN THE FORMATION AND ARRANGEMENT

OF HER PECULIARLY BEAUTIFUL

FLOWER GARDEN.

DESCRIPTION

OF

THE PLATES.

PLATE I.

PLAN of a FLOWER GARDEN *in the midst of Plea-sure Ground, surrounded by Shrubs.*

The borders may be easily arranged for the simple parterre. Forms 1, and 2, peculiarly adapted to the advantageous exhibition of flowers. General length of the beds from twenty-three to twenty-five feet. Width, in the broadest part, about four feet. Five or six feet of grass in the widest part between the beds, all the borders a good deal raised.

The tree at the entrance, which should be one of light, and rather pendulous foliage, must be cut to form a high stem, and the borders, if viewed under the branches, will have a beautiful effect. If the space of grass betwixt the borders appear too great, it may be lessened by baskets of ever-blowing roses, carnations, or any other plants; and these baskets may be formed by circular beds, surrounded by cast iron, made to resemble the open edges of a basket, and painted of a very dark green colour.

PLATE II.

PLAN of a FLOWER GARDEN upon a large scale, and more peculiarly adapted to the *Pleasure-Ground Garden,* although the form of the borders might be made use of in the common parterres, if judiciously planted so as to blend the variety of colours well with each other. The space of grass betwixt the shrubs and the borders should not be less than six feet.

Plate. I.

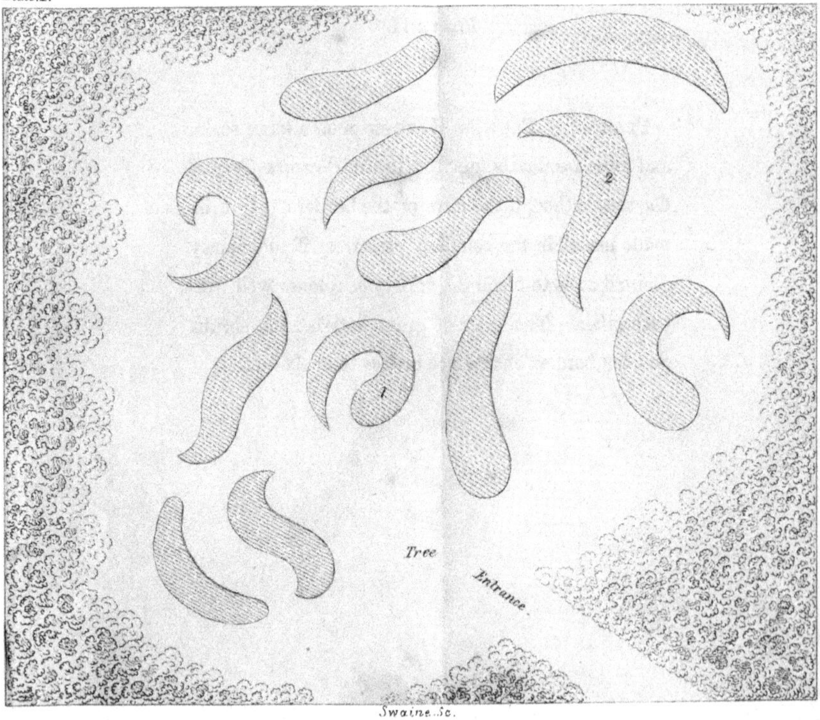
1
2
Tree
Entrance.

Swaine. Sc.

FLORIST'S MANUAL.

The material originally positioned here is too large for reproduction in this reissue. A PDF can be downloaded from the web address given on page iv of this book, by clicking on 'Resources Available'.

Plate II.

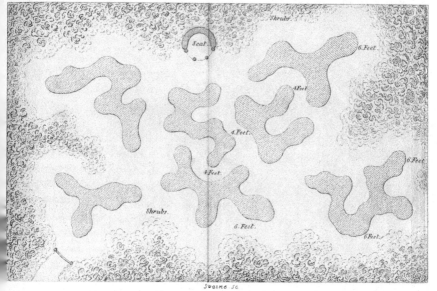

Swaine sc.

FLORIST'S MANUAL.

FLORIST'S MANUAL.

THE beautiful varieties of colour, form, and scent, exhibited in the structure of the vegetable creation, have, from the beginning of time, forcibly attracted the attention of mankind from the early age of infancy to the latest period of the decline of life; and have excited admiration from the inhabitant of the cottage, to him, the wisest of the human species, who dwelt in palaces, and spake of

B

plants, " from the cedar of Libanus, to
the hyssop which grew upon the wall."
We may then, perhaps, be allowed to
consider it as a part of the wisdom of
the present sapient æra, that the vegetable
species is become a subject of general
enquiry, and of prime consideration in
the arrangement of every modern dwel-
ling.

Omitting the scientific investigation
into the modes and habits of vegetable
existence, which affords a study of ex-
quisite delight to the ingeniously curious,
we confine ourselves to those gratifica-
tions only, which may be derived from
vegetables, to the visual, the olfactory,
and the saporific senses, their importance
to the latter being evinced by the expen-
sive buildings, extent of ground, and
numerous attendants appropriated to their

culture and accommodation, near all the habitations of the opulent; also, in every degree, from the luxurious exotic, fostered by the great, to the vine, which creeping around the cottage window, delights, at once, the eye, and gratifies the palate, of the humble inhabitant.

These grosser charms of vegetables, form, however, no part of our present enquiry. The universal taste, may it not be termed passion? now manifested for the accumulation and cultivation of flowers, is the main object of immediate consideration. Having, from early childhood to advanced age, possessed, I may almost say, an hereditary liking for this lovely order of creation, and having, from the subject, in all its branches, derived the most interesting amusement of my youth, I am solicitous to render

my sister florists partakers of my pleasures, so far, as by laying before them
a few hints, the result of experience, I
may enable them more methodically to
arrange their flowers, and so to blend
their colours, that through most part of
the spring and summer months they may
procure a succession of enamelled borders, which, without the knowledge of
the tints afforded by each season, cannot
be made to exhibit half the charms that a
flower-garden, well conducted, has the
capacity of presenting to the view.

It is to hints only that I pretend, nor
should I presume even so far, were I not
frequently consulted on the subject of procuring a GAY *Flower Garden*, and did I
not receive complaints from my florist
friends, that they find labour and expense
exerted in vain to the attainment of this

much desired object : labour and expense
will ever be in vain, unless the lady herself is capable of directing them to their
wished-for purpose, and it is to effect
this purpose that these few pages are
composed.

A Flower-Garden is now become a
necessary appendage of every fashionable residence, and hence it is more frequently left to the direction of a gardener,
than arranged by the guidance of genuine
taste in the owner; and the fashionable
novice, who has stored her borders, from
the catalogue of some celebrated name,
with variety of rare species, who has procured innumerable rose-trees, chiefly con-
sisting of old and common sorts, brought
into notice by new nomenclature, who
has set apart a portion of ground for American plants, and duly placed them in bog

soil, with their names painted on large headed pegs, becomes disappointed when, instead of the brilliant glow of her more humble neighbour's parterre, she finds her own distinguished only by paucity of colour, and fruitless expenditure.

Variety of species, bog borders, and largely lettered pegs, are all good in their way, but they will not produce a *gay* flower-garden ; and the simple cause of the general failure in this particular is the prevalent solicitude for rarity and variety, in preference to well-blended quantity ; as, without the frequent repetition of the same plant, it will be in vain to attempt a brilliant flower-garden, and, as in the judicious mixture of every common colour the art of procuring it consists. Hence, the foundation thus laid, the solicitude of those who wish to complete the superstructure must not be for rare species, but for new colour, so that the commonest primula which presents a fresh shade of red, blue, yellow, &c. ought to be esteemed more valuable than the most rare American plant which does not bring a similar advantage.

In the formation of that assemblage of flowers, which may be distinguished by the term of " The Mingled Flower Garden," it is essential that the separate parts should, in their appearance, constitute a WHOLE; and this appearance is more easily effected, if the borders are straight, and laid sideways, one before the other; but it is not incompatible with any form into which the ground may be thrown, if attention be given to the manner of planting.

In some gardens this appearance of a

whole is entirely destroyed by the injudi-
cious taste of setting apart distinct borders
for pinks, hepaticas, primulas, or any other
favourite kinds of flowers; also for different species of bulbs, as anemones, ranun-
culuses, hyacinths, &c.; these distinct borders, although beautiful in themselves,
break that whole which should always be
presented to the eye by the mingled
flower garden, as single beds, containing
one species only, form a blank before that
species produces its flowers, and a mass of
decaying leaves when the glow of their
petals is no more.

The reverse of this mode of planting is
essential to the perfection of the mingled
flower garden, in each border of which,
there should be, at least, two of every species; but the precise number must be regulated by the force of colour displayed by

the plant, and the size and the relative
position of the borders. It will be only
necessary to observe that, to whatever view
the garden presents itself, the eye should
not be checked by the failure, in any
part of it, of the prevalent colours of the
season. The situation of a flower garden
is rarely left to the free option of the
owner, that option being generally con-
trolled by a variety of small circumstances to which she will, in some degree,
be obliged to submit, and more particularly so in that humble flower garden for
the construction of which alone, I pretend
to offer hints of direction; but this, although the one most easily to be obtained,
should not be neglected, even by those
who have the power of cultivating exotics
in their highest perfection.

The　common, or Mingled　Flower

Garden should be situated so as to form an ornamental appendage to the house, and where the plan of ground will admit, placed before windows exposed to a southern or south-east aspect; and, although, to this position there may appear the objection of the flowers turning their petals to the sun, and consequently from the windows, this predilection in the tribe of Flora for the rays of that bright luminary, will produce the same effect in whatever place our flowers may be situated, when in the vicinity of a building, as they invariably expose the front of their corols to the light, from which both the petals of flowers and the leaves of plants are believed to derive some material essential to their existence.

The compass of ground appropriated

to flowers must vary according to the size of the place of which that ground forms a part, and should in no case be of great extent. The principle on which the parterre should be laid out, ought to be that of exhibiting a variety of colour and form so nicely blended as to present one whole. In a flower garden viewed from the windows of a house, this effect, as has been observed above, is best produced by strait borders laid sideways of each other, and to the windows from whence they are seen, as by that position the colours shew themselves in one mass, whereas, if placed end-way, the alleys, which are necessary for the purpose of going amongst the flowers, divide the whole, and occasion an appearance of poverty. Should an intermixture of turf with the flower borders be preferred, then the borders should be of various forms, examples of which are prefixed to the book*.

It is more difficult than may at first appear, to plan, even upon a small scale, such a piece of ground, nor perhaps, would any but an experienced scientific eye be aware of the difficulties to be encountered in the disposal of a few shaped borders interspersed with turf; the nicety consists in arranging the different parts so as to form a connected glow of colour, to effect which it will be necessary to place the borders in such a manner that when viewed from the windows of the house, or from the principal entrance into the garden, one border shall not intercept the beauties of

* See Plates.

another, nor in avoiding that error, pro
duce one still greater, that of vacancies
betwixt the borders forming small avenues,
by which the whole is separated into
broken parts, and the general effect lost.

Another point to be attended to is
the just proportion of green turf, which,
without nice observation, will be too much
or too little for the colour with which it is
blended; and lastly, the breadth of the
flower borders should not be greater
than what will place the roots within
reach of the gardener's arm without the
necessity of treading upon the soil, the
mark of footsteps being a deformity
wherever it appears amongst our flowers.
If the form of ground where a parterre
is to be situated is sloping, the size
should be larger than when a flat surface,
and the borders of various shapes and on

a bolder scale, and intermingled with grass; but such a flower garden partakes more of the nature of pleasure ground than of the common parterre, and will admit of a judicious introduction of flowering shrubs.

If it happen that a house be nearly surrounded by a flower-garden, the variety of aspect thence afforded will be favourable to the continuance of the bloom of our flowers far beyond what can be obtained if confined to a southern exposure. South, south-east, and east, are the aspects most advantageous to the growth of flowers; and, possessing these varieties of exposure, the bloom of a garden may be protracted some weeks beyond the time it could be preserved under a single aspect. When apart from the house, the Mingled Flower

Garden may be introduced with great
advantage, if situated so as to form a
portion of the pleasure-ground: in this
case it should not be distant from the
house, but so contrived as to terminate
one of the walks of the home shrubberies;
the garden must be situated south, or
south-east, and the fence, which will be
necessary for protection from hares and
other animals, should be made of wire,
and, in some peculiar situations, might,
perhaps, be nicely hidden by low shrubs,
periwinkle and other running plants,
which will readily grow upon mossy
trunks, roots, or arms of old trees: and
these, thrown carelessly on the ground,
and judiciously planted, might form a part
of the beauty of the garden, while they
served the purpose of veiling the fence
from the eye; also, fragments of stone

may be made use of, planted with such roots as flourish among rocks, and to which it might not be difficult to give a natural appearance, so far as by bringing forward to the view the utility of these stones in the culture of the vegetables growing thereon, while the real purpose of breaking the line and concealing the boundary fence might be disguised.

The present fashion of introducing into flower-gardens this kind of rock-work requires the hand of taste to assimilate it to our flower borders, the massive fabric of the rock being liable to render the lighter assemblage of the borders diminutive and meagre: on this point, caution only can be given, the execution must be left to the elegant eye of taste, which, thus warned, will quickly perceive such deformity.

I must venture to disapprove the extended manner in which this vegetable rock-work is sometimes introduced, not having been able to reconcile my eye, even in gardens planned and cultivated with every advantage which elegant ingenuity can give them, to the unnatural appearance of artificial crags of rock and other stones interspersed with delicate plants, to the culture of which the fertile and sheltered border is evidently necessary, being decided that nothing of the kind should be admitted into the simple parterre that is not manifestly of use to the growth of some of the species therein exhibited.

In pleasure-grounds or flower-gardens on an extensive scale, where we meet with fountains and statuary, the greater kinds of vegetable rock-work might pro-

bably be well introduced; but to such a
magnificent display of art I feel my taste
and knowledge wholly incompetent.   I
attempt only to assist in the humble path
of exhibiting to the best advantage the mo-
derately-sized flower-garden, replete with
colour of every variety, and in order
to the procuring such variety I shall an-
nex to this little book a short list of the
commonest plants which expand their
beauties at the same season, and of the
colours prevalent in that season, so that
by consulting that list any one may be
enabled to form a gay and well-mingled
garden throughout the spring and summer
months at a small expense; and thus,
having formed the basis, more rare plants,
or a more extended variety may be su-
peradded, as choice or circumstances
may admit.   Also, where neither expense

nor trouble oppose their prohibitory bar-
rier, many of the vegetable tribe may be
cultivated to greater perfection, if we
appropriate different gardens to the
growth of different species, as, although
it is essential to the completion of our
first kind of garden to introduce, on
account of their scent and beauty, some
of the more hardy species of the flowers
termed annuals, in that situation room
cannot be afforded them sufficient to their
production in that full luxuriancy which
they will exhibit when not crowded and
overshadowed by herbaceous vegetables;
and hence becomes desirable that which
may be called The Annual Flower Garden, into which no other kind of flower
is admitted besides that fugacious order,
and under which is contained so great a
variety of beauty and elegance as one

well calculated to form a garden, vying in brilliancy with the finest collection of hardy perennials.

Also, the plants comprised under the bulbous division of vegetables, although equally essential to the perfection of The Mingled Flower Garden, lose much of their peculiar beauty when not cultivated by themselves, and will well repay the trouble of an assiduous care to give to each species the soil and aspect best suited to its nature. Two kinds of garden may be formed from the extensive and beautiful variety of bulbous-rooted flowers, the first, wherein they should be planted in distinct compartments, each kind having a border appropriated to itself, thus forming, in the Eastern taste, not only the " garden of hyacinths," but a garden of each species of bulb which is capable of being brought

to perfection without the fostering shelter
of a conservatory. The second bulbous
garden might be formed from a collection
of the almost infinite variety of this lovely
tribe, the intermixture of which might produce the most beautiful effect, and a succession of bloom to continue throughout
the early months of summer. A similar
extension of pleasure might be derived
from a similar division of all kinds of
flowers, and here the taste for borders
planted with distinct tribes may be properly exercised, and, as most of the kinds
of bulbs best suited to this disposition have
finished their bloom before the usual time
at which annuals disclose their beauties,
the annual and the bulbous gardens might
be so united, that, at the period when the
bloom of the latter has disappeared, the
opening cords of the former might supply

its place and continue the gaiety of the
borders; nor is there the same inconvenience in planting together annuals and
bulbous roots, as when annuals are mingled with a mass of herbaceous plants, the
leaves of the bulbs being past their period
of growth, and on the decline, may be
tied together without the hazard of injury
to the *forming bulb, and thus kept from
over-shadowing the tender growing plants
of the annuals.   The ingenious Florist
will perceive that by the skilful conduct
of separating and combining, she may

* As all bulbs are annually renewed by the growth of
a new bulb formed and nourished from the bulb of the
preceding year and from the recrements of its foliage,
many bulbous plants are destroyed, or materially weakened by the ignorant practice of cutting off the leaves as
soon as the flowers are faded.—See Sketches of the
Physiology of Vegetable Life, page 156, and plate 12.

multiply and vary the display of her
flowers to the utmost extent that her
fancy may suggest; but in such a fan-
tastic extent of her power I do not pretend to accompany her, nor even to offer
directions for any kind of garden except
that which may be generally attainable.
I must, however, recommend a spring
conservatory, annexed to the house, con-
sisting of borders sheltered by glass and
heated only to the degree that will produce a temperate climate, under which all
the flowers that would naturally bloom
betwixt the months of February and
May, might be collected, and thence be
enabled to expand their beauties with
vigour, which, when they are exposed to
the vicissitudes of the open air, becomes
so impaired by the harsh winds of spring
as annually to blight their charms, and

disappoint our expectations; so that we usually think ourselves fortunate if we are able to preserve the roots alive, encouraging ourselves with the hope of the future year, which hope is again disappointed as spring with its chilling blasts returns.

Weather, however, is not the only enemy from which we have to fear the destruction of our plants; insects of all kinds and degrees attack our seeds, our roots, and our flowers: hence directions for the prevention of such depredators become a necessary part of a work which has for its object the exhibition of the floral world to its greatest advantage, and as amongst the various receipts given by all gardeners for the destruction of insects, I have not found any which can be esteemed efficacious, I hope I may not

appear too diffuse in my detail of the only
method which, I believe, will clear our
borders of these enemies, and which, if
skilfully followed, may nearly effect their
annihilation.

The simple and laborious mode of pick-
ing away the animal, is the only one to
which recourse can be had with permanent advantage; and to give full efficacy
to this method of rescuing our plants
from caterpillars, snails, &c. our attacks
must be made upon them at particular
seasons, and a knowledge acquired of
their history, so far as to enable us to
have swarms of them destroyed in the
destruction of an individual of the species;
without, however, much research into their
natural history we may, from common
observation, understand that in the winged
insect we may free our plants from an

innumerable tribe of those which crawl, and which, in that reptile state, have the capacity of devouring the whole product of a garden.

The two periods of change of form in the caterpillar species seem to afford the most advantageous times to put an end to their existence, as in the ephemeral butterfly, if timely attended to, we may destroy the animal before it has acquired the power of disseminating its young progeny; and, in the intermediate and voracious state of caterpillar, every single one which is prevented attaining the winged form preserves our flowers from an host of enemies.

The green caterpillar is the most common foe to our flower borders, and in autumn attacks the branches of mignonette in such numbers, as to afford an easy opportunity of their destruction. A more persevering enemy, and one more difficult to exterminate from gardens, is the snail, or common slug, which, forming its habitation under the soil, attacks the roots of flowers, and frequently destroys them, before the gardener can be aware of the mischief, that too often becoming visible only when past reparation. Under a vigilant eye, however, plants will not twice suffer from the enemy not being ostensible; as the symptoms of his vicinity may be marked by flowers perishing as they first emerge from their buds or bulbs, by leaves or petals being pierced in small holes, or having the appearance of being gnawed, or in growth, or from, almost, any failure in vigour which cannot be accounted for by external causes.

In my early acquaintance with the pernicious effects of snails, having observed a root of hepatica, which had been recently planted, fade and shew symptoms of some fatal malady, I caused it to be taken out of the ground, and found amongst the fibres of its roots a number of those beautiful pearl-like substances, which are the eggs of the snail. Having caused these, with some snails, which were also found amongst the roots, to be taken away, and the hepatica to be re-planted, I soon perceived the good effects of having dislodged the enemy, as the plant flourished from that period.

In cold and dry weather the snail rarely appears, but after warm showers it may generally be found; early in the morning, and about the close of evening, are the usual times of these insects coming abroad, when they may be picked up in large quantities. They will, however, frequently molest a plant for a length of time, without being visible, in which case, when there is reason to suspect the hidden attacks of snails, the only method to entrap them is to place a common garden-pot over the infested root, and it will rarely occur that the enemy is not discovered, as snails fasten themselves to the sides or tops of pots, boards, or mats so placed, and, thence, are easily taken. In droughty, seasons it will be of use to water the plant before it is covered, as the moisture of the earth will be an additional motive of attraction to draw the animal from his hiding place.

And here I must be allowed to recommend to all those, who, for the protection of their flowers, and fruits, are obliged to destroy an order of creation, indubitably

endowed with sensations of pleasure and
pain, to take care that their existence is
put an end to with humanity ; if thrown
immediately into water, the snail is
instantly destroyed, and consequently can
scarcely be susceptible of suffering.

The smaller insects which infest rose-
trees, and some herbaceous plants, can
only be kept within moderate bounds by
sweeping them from the branches, or by
cutting off those whereon they are found
in most profusion.—In carrying off these
diminutive enemies, birds are peculiarly
serviceable; and a well-authenticated fact,
which I have received, of the conduct
of a hen with her chickens, seems to hint
that we might render them of use in our
gardens, although it may be doubtful
whether the injury liable to be sustained
by the scratching of their claws, would

not counterbalance the advantage of the number of insects cleared away by their beaks.—The fact was stated to me as follows.

A lady, whose garden was enclosed by a hedge of rose-trees, and which rose-trees were covered by swarms of minute insects, saw a hen lead her flock of chickens into the garden; her immediate intention was to have them driven out, but she soon perceived their eyes fixed upon the rose-trees, and watched them until they had satiated their appetites, and perfectly cleared some of the trees.

In the attention given to the habits of snails it should be peculiarly exerted at the time when a plant is first put into the ground, and again when it puts forth its vernal buds, also when, after having flowered, the leaves begin to decay, at

which period bulbs are apt to be lost, and
most frequently, in consequence of the attacks of snails, as at that time they are
not only infested by the snails of complete
growth, but also with numbers recently
come forth from their eggs, and of a size
scarcely equalling that of the head of a
large pin, and these minute animals, if
not destroyed, will deprive many bulbs,
and also many buds of herbaceous plants
of their existence.

It is remarkable that insects generally
attack those plants which are least vigorous, and the reason of their selection of
such leaves as are beginning to decay may
be, that in their declining state they have
usually a peculiar sweetness, probably,
perhaps owing to some saccharine juices
which are preparing for the nutriment of
the bulb or bud which is forming in their

bosoms, it being known to botanic philo-
sophers that the nascent vegetable derives
its sustenance from the recrements of the
one from which it takes its birth.

———————

And now, trusting that the hints contained in these few pages may enable my
sister gardeners to cultivate their flowers
to a degree of perfection suited to their
wishes, and, by so doing, render them objects of their genuine admiration, I will
not disguise my earnest desire to lead them
from the pleasure they receive in the superficial view of a profusion of gay and
varied colours before their windows, to
the investigation of the habits and pro-

perties of these elegant playthings, as in every change of season, amusement, ever new and varying, may be derived from the study of vegetable existence.

The dreary months of winter, which, to the uninformed eye, exhibit only destruction and desolation, present to that of the botanic philosopher a scene of order, renovation, and beauty, while he contemplates the infinite variety which forms the whole of that vast plan of care and preservation evinced in the mechanism of the minutest bud, which awaits only the genial breath of spring to expand its wonders to the day.

In the slow and gradual decay of the foliage of his trees, he sees, from the recrements of that foliage, an increase as slow and gradual of the buds which are preparing, in their turn, to enjoy the

transient pleasures of existence; and as
the leaves of the flower borders fade
away, and, apparently, perish, the philosophical florist perceives, in their decay,
new birth given to a viviparous progeny,
with the same certainty as the seed buried
within the earth reproduces its seminal
posterity, or as the butterfly arises from
its chrysalis.

I hope I shall not be deemed presump-
tuous in recommending to the perusal of
genuine Florists, a small tract, entitled,
*Sketches of the Physiology of Vegetable
Life,** which, being chiefly the result of
simple experiments, is calculated to in-
struct those young persons, who, while
they amuse themselves by the culture of
their gardens, may not have either leisure

* Sold by Hatchard, Piccadilly.

or inclination for actual study, and may be pleased to find collected, in a few pages, a variety of interesting and highly curious facts relating to the cherished objects of their attention, and which may be understood without the labour of close application. Therein, also, the young Florist will find a view of the wonderful process which takes place in the reproduction of all bulbs, the knowledge of which may be esteemed essential to the conduct of their increase, and which ought to be acquired by all who are desirous of possessing, in perfection, those prime treasures of the floral amateur.

In having condemned the search after rarity and variety, I must be understood to confine my disapprobation of this pursuit to the general Florist only: to the classical botanist variety and rarity are of the first value; hence the gardens of the classical botanist and general florist differ, even in their first principles. The botanist will justly estimate the value of her garden by the number of genera, and the variety and rarity of species therein collected; and while, to the comprehension of the Florist, there is little exhibited besides the lettered pegs which obscure, while they enumerate, the plants, the classical botanist will exult in the possession of a greater number of species of some rare individual genus than, perhaps, it may be within the power of botanists, in general, to obtain.

The botanist, and the general Florist, for I speak not of those Florists who confine their admiration of flowers to the greater or lesser number of stripes in the petals of a tulip or of a carnation, are

more nearly allied in their tastes than
may, at first, appear. That which pleases
one, gratifies the other; and it is only in
the extent of their observation that they
will be found to differ. The sleep of
plants, their various modes of inflorescence, the annual phenomenon of germination, the change of position of the seed-
vessels, through the marvellous process of
fructification, have each excited the sur-
prise and admiration of every intelligent
Florist. She observes, and is amused by
such appearances, but exerts her intellect
no farther; while the philosophic botanist
reasons from effect to cause, until she
cannot refuse her belief that the curious
and beautiful economy of vegetable existence must proceed from laws not purely
mechanical. Notwithstanding the distinction we find between the classical and

the philosophical botanist, and yet greater
betwixt the scientific pursuit of the knowledge of flowers, and that of merely
arranging them into an assemblage of
colours, I venture to assert that, while it
is essential to the botanical philosopher to
be acquainted with an accurate view of
the science of classification, the Florist
will increase her amusement ten-fold by
making herself familiar with the ingenious
system of the great parent of botany,
Linneus, and some knowledge of which
seems unavoidable in those ladies who,
in cultivating their favourite flowers, exercise the mental along with the corporeal
faculty.

It is certain, however, that an inquiry
into the science of the subject is by no
means essential to the pleasure which may
be derived from the culture of a flower-

garden; and, notwithstanding that I recommend to the genuine Florist a more
extended acquaintance with the economy
and habits of the vegetable tribe, the
wonders of which are hourly passing
before her eyes, I have too much experience of the delight which may be excited
by the bare view of the simplest flower of
our meadows, or of our hedge-bank, to
entertain a doubt of the gratification received by the general Florist from the
superficial contemplation of her cultivated
borders.   I shall, however, esteem myself
happy if by these trivial observations I induce, even a few, of my sister Florists to
exercise their intellect, or relieve their
ennui by an inquiry into the causes whence
those effects proceed, which, while gather-
ing a common nosegay, cannot but frequently have solicited their attention.

Nor is it only the amusement of the present moment that I seek to afford. To use and not to fatigue the understanding, to interest and not to absorb the mind, is the true art by which happiness is to be attained; and, while from the wonderful structure of the creature, we are led to the contemplation of the Creator, we shall find this a more certain panacea to the daily chagrins of human life, than all that the dissipation of the gilded hours of indiscriminate society has ever been able to afford.

M. E. J.

Somersal Hall.

A

CATALOGUE

OF

COMMON HERBACEOUS PLANTS,

&c.

CATALOGUE

OF

COMMON HERBACEOUS PLANTS,

With their Colours, as they appear in each Season
from FEBRUARY to AUGUST.

The Names of the Flowers accented according to the
Lichfield Translation of the System of Vegetables
of Linneus.

V. marks *varieties,* of a true species.

FEBRUARY.				MAY.
		RED.		
Anémone.	.			
hepática,	.	.		Single and double.
horténsis,	.	.	.	Varieties.
Alýssum,	.	.	.	.	Alysson.
deltoídeum	.	.	.	Purple

RED.

Béllis, ...   Daisie.

         perénnis,   .   V. from deep crimson to
                                      pink and white.

Erínus, ...   Grows low.

         alpínus,   .   .   .   Pretty.

Erythrónium dens canis,   .   Dog's tooth violet.

Fritillaria imperiális,   .   .   Crown imperial.

         meleagris,   .   .   .   With Vs.

Fumária, ...   Fumitory.

         solida,   .   Bulb-rooted, flowers early,
                                      troublesome, from seed-
                                      ing profusely.

Hyacínthus orientale,   .   Oriental, single and double.

Oróbus vérnus   .   .   .   Spring vetch.

Phlón, Lychnidéa.

         subuláta, ...   Awl-shaped.

         setacea,   .   .   .   Bristly.

Prímula vulgáris,   .   Common primrose.   Vs. in
                                      shades of red, single and
                                      double, including double
                                      Polyanthus, which gives

FEBRUARY.   MAY.

RED.

	a handsome very dark shade of red.
Villósa, . .	Villous, beautiful, and V.
longiflóra, . .	Long-leaved.
farinósa, ...	Mealy.

BLUE.

Anémone hepática, .	Single, semi-double, and double.
pulsatilla,	
apennina, ...	Apennine.
Cynoglóssum, . . .	Hound's tongue.
omphalódes, .	Comfrey-leaved.
Crócus,	
vérnus, ...	Spring.
Hyacínthus,	
botryoídes, . . .	Grape.
comósus, . . .	Purple grape.
Iris,	
púmila, ...	Dwarf.
Prímula, . .	Auricula, deep blue, with the eye brimstone-coloured.

FEBRUARY.   MAY.

BLUE.

Pulmonária, ...   Lung-wort.

    officinális,   .   .   .   Officinal.

    Virgínica,   .   Virginian, bright blue.

Scílla,

   præcox,   .   .   .   Early-flowering.

   bifólia,   .   .   .   Two-leaved.

   vérna, ...   Vernal.

     All pretty, grow low ;— many bulbs should be planted together.

Vióla,   .   .   .   Pansie, tri-colour, very large, rich blue ; and paler shade, with the flowers of less size.— V.

YELLOW.

Adónis vernális,   .   .   .   Spring Adonis.

Alýssum,   .   .   .   Alysson of Crete.

   saxátile, ...   Rock.

   mínimum,   .   .   .   Smallest.

Crocus,

   vérnus,   .   .   Spring.

FEBRUARY.   MAY.

YELLOW.

Crócus,

     sulphureus,   .   .   .   Sulphur.

     susiánus,   .   .   Cloth of gold.

Erythrónium, ...   Dog's tooth.

     Americánum, American. Not so hand-
some as the other spe-
cies.

Fritillária imperiális,   .   .   Crown imperial.

Helléborus hyemális,   .   .   Winter aconite.

Narcíssus,

     angustissimus,   .   Narrow-leaved.

     minor,   .   Grows low ; very pretty.

     bulboiódium,   .   Hoop-petticoat.

     . *N.* Grows low, and gives
a deep shade of yellow.

     triandrus,   .   Pale yellow, 3-stamened,
very pretty.

Narcíssus,

     jonquilla,   .   Jonquil, single and double.

Pseudo-Narcissus,   .   Daffodil with Vs.

YELLOW.

Bícolor,   .   Butter and eggs, single and double.

Taretta,   .   .   .   Polianthus with Vs.

    prímula,   .   Auricula, single and double, the double beautiful.

Veris,   .   .   .   V. ox-lip and cow's-lip.

.

WHITE.

Anémone, ...   Wood.

    nemorósa,   .   .   Single and double.

    hepática,   .   More rare and more tender than the coloured.

Arabis alpína,   .   .   .   Wall-cress alpine.

Béllis, ...   Daisie.

    pcrénnis,   .   .   V. double, very pretty.

Cardámine praténsis,   .   Lady's smock, double.

Crocus, ...   Scotch crocus.

    biflórus,   .   Two-flowered. Valuable for blowing some weeks before crocus vernus.

FEBRUARY.   MAY.

WHITE.

Erythrónium,

dens-canis,   Dog-tooth. More rare than the red, a beautiful feature in the mingled flower-garden : not less than ten bulbs should be planted together.

Galánthus nivalis,   .   Snow-drop, single and double.

Helléborus niger,   .   .   .   Christmas rose.

Lecucójum,   .   .   .   Snow-flake.

vernum, ...   Spring.

Prímula nivalis,   .   .   .   White auricula.

vulgáris,   .   V. paper-primrose, single and double, hose in hose.

Ranúnculus amplexicáulis, Stem-clasping plantain-leaved crow-foot.

Sanguinária Canadense,   .   Puccoon, Canadian.

Tiarélla cordifólia,   .   .   Heart-leaved.

D 2

MAY.   AUGUST.

RED.

Antirrhinum,   .   Snap   dragons;   various
                         shades.

Astrantia,
    major,
    minor.

Aquilégia,   .   .   .   Colombine.
    vulgáris,   .   Common; many varieties,
                         the starry very pretty.

Canadénse,   .   .   Canada; red and yellow.
Anémone,
    horténsis,   .   Many Vs; from deep scar-
                         let to pink and white.
                         By sowing seed every
                         spring, and planting the
                         roots at different periods,
                         the bloom of this beau-
                         tiful flower may be con-
                         tinued through most part
                         of the spring and sum-
                         mer months.

Béllis prolífera,   Hen and chicken daisie.

MAY.   AUGUST.

RED.

Cistus,

  heliánthemum,   .   .   .   Dwarf.

Cheiránthus, cheiri,   .   Bleeding wall-flower.

  annuus,   .   .   .   Stock, ten weeks.

  incanus,   .   .   .   Brompton stock.

Chelóne,

  barbáta,   .   .   Beard-flowered.

  obliqua,   .   .   Red-flowered.

Diánthus, ...   Sweet-William.

  barbátus,   .   .   Mule and tree-mule.

  superbus,   .   .   .   Superb.

  cœsius,   .   Mountain : star-pinks and

        variety of carnations.

Dictámnus, ...   Fraxinella.

Dodecátheon,

  meadia,   .   .   Virginian.

Epilóbium,   .   .   .   Willow-herb.

  angustíssimum,   .   Rosemary-leaved.

Fumária,   .   .   .   Fumitory.

  formósa,   .   .   Red-flowered.

RED.

Geranium macrohirum,	. .	Long-rooted.
sylváticum,	. .	Wood.
sanguineum,	. .	Bloody.
Lancastriénse,	. .	Lancashire.
Gladiólus commúnis,		Corn-flag common.
Iris versicolor,	. . .	Various-coloured.
Láthyrus latifólius,	. .	Everlasting pea.
Lílium chalcedónicum,	. .	Martagon, scarlet.
Lýchnis,	.	
viscária,	. . .	Viscid.
flos-cucúli,	.	Ragged robin, double.
chalcedónica,	.	Scarlet, single and double.
Lýthrum,		
Salicária,	. . .	Common.
virgátum,	. . . .	Twiggy.
Monárda dídyma,		Common scarlet and pale purple.
Orobus várius,	. .	Red and yellow vetch.
Orchis máscula,	.	Deep shade of purple red: very good effect. See observations, p. 69.

RED.

Papáver, ...   Poppy.

    orientale, ...   Eastern.

Pæónia, ...   Peony,

    officinalis,   .   Common, dark, double red, and rose-coloured.

    tenuifólia,   .   Fine-leaved Lychnidéa.

Phlon,

    glaberrima,   .   .   Smoothest.

    stolonífera, ...   Creeping.

    ováta, ...   Oval-leaved.

    amœna,   .   .   .   Fine red.

    intermedia,   .   .   .   Intermediate.

    pilósa, ...   Very pretty.

    maculáta,   .   .   .   Spotted.

Rudbeckia purpurea,   .   .   .   Purple.

Scilla, ...   Hare-bell.

    nutans,   .   .   Flesh-coloured.

Tulipa gesneriána,   .   Garden tulip. Single and double; single, rich deep red: very good effect.

RED.

Túlipa, . .		Tulip, dwarf.
suavéolens, .		Van Tol. sweet scented.
Cleremont, . .		Pink and white.
Thalictrum aquilegifólium, .		Meadow rue.
		Columbine - leaved, with purple flowers.
Valeriána, ...		Valerian.
rubra, . .		Red, two shades.
Verónica, ...		Spiked.
carnea, .		Flesh-coloured, two shades.

BLUE.

Anemone,

horténsis, V. . The double kinds, not adapted to mingled flower borders, as they require peculiar culture to bring them to perfection.

Aster, alpínus, . . Handsome grows low.

MAY. AUGUST.

BLUE.

Aconitum, . . . Monk's-head.

 napellus, . . V. Blue and white.

Campánula,

  persicifolia, Peach-leaved, single and double.

  pumila, . . . Dwarf.

  carpática, . . Carpathian.

Catandnche cærúlea.

Cheiránthus, incanus an-

 nuus, . . Brompton stock, 10 weeks. By sowing the seed of stocks, and putting out the plants at different times, the bloom may be continued until destroyed by frosts.

Centauréa, cйanus, . Corn-bottle; large flower; fine bright deep blue; not in esteem with florists, but worthy of a place in the Mingled Flower-Garden.

MAY.   AUGUST.

BLUE.

Delphinium,	. . .	Larkspur.
grandiflórum,		
elátum,	. . . .	Bee.
azureum,	. . . .	Azure.
Gentiána,	. . . . .	Gentian.
sapondria,	. . .	Soap-wort.
septemfida, asclepiadea		7-cleft swallow-wort.
acaulis,	.	Gentianella; the last spe-

cies, planted at the edge
of a border, facing the
south, in a row of five
or six inches broad,
makes a superb appear-
ance.

Gerànium, palustris,	. .	Single and double.
Hemerocállis,	. . .	Day-lily.
cerulea,	. .	Blue flowered.
Iris,		
cristáta,	. . .	Crested.
sambucina,	. .	Deep blue.
German,	. .	Pale blue, beautiful.
xiphium, xiphioides,		Small and great bulbous.

MAY. AUGUST.

BLUE.

Linum, ... Flax.

    perenne, . . . Perennial.

    pumila, . Dwarf. — Marked annual
in Mr. Donn's catalogue;
certainly continues more
than one year.

Lupinus, ... Lupine.

    perennis, . Perennial, two kinds.

    polemonium cæruleum, . Greek valerian.

Phytéuma, . . . Bright deep blue.

    orbiculáre, . . Round-headed.

Scilla,

    campanuláta, . . Bell-flowered.

    nutans, . . . Hare-bell.

Sophóra austrális, . . . Blue-flowered.

Verónica,

    prostráta, ... Trailing.

    chameedry, . . Germander.

    incana, . . . Hoary.

    spicáta, . . . Spiked.

    gentianoídes, . Gentian-leaved : pale blue,

BLUE.

of a shade very uncom-
mon ; very good effect.

Vinca major,   .   .   .   Periwinkle.

When the trailing branches
are cut off, the vinca major
with its varieties, will
grow in small bushes, and
is pretty ; in its natural
trailing state it is very
ornamental amongst rock
work.

YELLOW.

Allium moly,   Disagreeable, — from   its
strong onion smell ; valu-
able as it supplies a shade
of deep yellow, late in
June.

Antirrhinum spartium, . Annual   broom. — Grows
very low, and should
be   sowed   near   the

YELLOW.

		edges of the borders; essential to the beauty of mingled flower gardens, from June to September.
Cáltha palustres,	.	.	Meadow bout, double.
Cheiránthus cheiri,	.	Green-top, or yellow wall-flower, double.
Cistus heliánthemum,		.	.	Dwarf.
Coreópsis,	.	.	Tick-seed sun-flower.
tenuifòlia,	.	.	.	Slender-leaved.
aúrea,	.	.	.	Golden.
verticilláta,	.	.	Whorl-leaved.
Hemerocállis,	.	.	.	Day lily.
flava,	.	.	Yellow.
fulva,	.	.	Copper-coloured.
Lilium,			
Canadénse,	.	.	Turk's cap.
bulbíferum,	.	.	Bulb-bearing.
tigrínum,	.	.	Tiger-spotted.
Ænothera,	.	.	.	Tree-primrose.
pumila,	.		Dwarf; very low.

MAY.   AUGUST.

YELLOW.

fruticósa,   .   .   Perennial.

Papáver, ...   Poppy.

Cámbricum,   .   .   Welch; perennial.

Túlipa,   .   .   .   Dwarf; very pretty.

sylvestris,   .   Single; flowers nodding;
blows early.

Gesneriána,   .   .   V. Double yellow.

Tróllius, ...   Globe.

Europaius,   .   .   European.

Asiáticus,   .   Asiatic; colour of Asiati-
cus, peculiarly good ef-
fect.

Vióla, ...   Pansie.

tricolor,   .   .   .   Varieties.

grandiflóra,

lutéa,   .   .   .   Yellow.

MAY.   AUGUST.

WHITE.

Antirrhinum, . . .	Snap-dragon.
Anthéricum,	
liliágo, . .	Grass-leaved.
liliástrum, . .	Savoy spider-wort.
Actæa racemósa, . . .	Branched.
Anémone, ...	Snow-drop leaved.
dichótoma, . .	Two-forked.
Bellis, ...	Daisie.
perénnis, . .	Double, very pretty.
Campánula persicifólia, .	Peach-leaved, single and double.
púmila, ...	Dwarf.
Cheiránthus, ...	Stock.
incánus, . . .	Brompton.
ánnuus, . .	Ten-weeks.
Convallária polygonátum,	Solomon's seal, single and double.
Dictámnus, ...	Fraxinella.
Hésperis matronales, . .	Rocket, double.
Iris, ...	Large.
xiphoides, . . .	Bulbous.

WHITE.

Lilium, ...   Lily.

        candidum,   .   .   .   White.

Narcissus,

        poéticus,   .   .   Poet's; double.

Ornithogalum,

                pyramidale,   .   .   Pyramidal.

Phlox,

        suavéolens,   .   .   Sweet-scented.

Pancrátium,

                maritinum,   .   .   Sea.

Polýgonum,

                viviparum,   .   Viviparous; grows very
                        low, pretty.

Ranúnculus,

                aconilifolius,   .   Mountain; double.

Saxifraga,   .   .   .   Double.

                granuláta,   .   Grain-rooted; very orna-
                        mental before flowering
                        by the green patches of
                        the foliage amongst the
                        early spring flowers.

WHITE.

Scilla,

    campánulata, . . Bell-flowered.

    nútans, . . . Hare-bell.

Stipa, ... Feather grass.

    pennáta, ... Soft.

Spiríea,

    aruncus,

    filipéndula, . Drop-wort; double.

    ulmária, . Meadow-sweet; double; the single kinds have little beauty.

    trifoliáta, . . Three-leaved.

Thalíctrum,

    aquilegifólium, . Colombine-leaved.

Túlipa, ... Tulip.

    gesneriána, . V. Slightly streaked with pink.

Verónica,

    spicáta, . . . Spiked.

    pinnáta, . Pinnate; the prettiest of the spiked veronicas.

MAY.   AUGUST.

WHITE.

Vinca, minor   .   .   Periwinkle, with variegated
                             leaves, very pretty when
                             cut into bushes.

OBSERVATIONS.

MANY flowers in the foregoing catalogue continue in bloom from July to October; the herbaceous plants, which flower in autumn, are generally large, some of them. extremely handsome, and, in extensive flower-gardens, produce a very ornamental effect; álcea rosea, hollyhock, with all its beautiful varieties, many species of perennial ash, the dáhlias, échinops, sphærocephalus, globe thistle, the common sun-flower, and some other species of heliánthus, will not escape the attention of the genuine Florist, if the compass of her ground be large enough to admit of their introduction. For the common-sized Mingled Flower-Garden, from the beginning of August, the chief dependence for gaiety must be upon annuals, the hardy kinds of which are so generally known, as to render unnecessary the enumeration of them in this place. Carnations also may contribute their share to the brilliancy of our autumnal borders; but there are few plants so ornamental, at that season, as the double dwarf poppy, which displays an endless variety of colour in the shades of red, and also produces perfectly white flowers, with the petals of a most delicate texture; the seed of these poppies should be scattered all over the borders and suffered to grow promiscuously, as chance may direct, only taking out a few of the plants, where they grow too

thick. China-asters and marigolds may
be planted in the borders near the
patches of crocuses and snowdrops, the
leaves of which have disappeared. The
single and double colchicums are beautiful, and give gaiety to our gardens at a
late season. The popular belief, that the
fruit, or seed of the colchicum, is produced previously to the flower, is wholly
unfounded, and, as the peculiarities in the
appearance of the fructification of this
plant generally excites the curiosity of
Florists, I venture to refer the ingeniously
inquisitive to " Physiological Sketches
V egetable Life," page 160, plate XI.
where they will find full information on
that interesting subject. The orchis mas-
cula, which from the rich purple of its
petals, and dark-spotted leaves, merits
a place amongst our cultivated flowers,

is rarely seen in gardens, it being generally supposed that there is some peculiar
difficulty in removing the roots of this
curious tribe of plants from their native
situations of growth. I have in a former
work* hazarded the conjecture, that the
orchis, in removal, did not require different treatment from that necessary to
be given to all other bulbous plants under
the same circumstances; and I have since
confirmed the justness of this conjecture
by experiment. It is requisite that the
leaves of all bulbous plants should be
wholly decayed before their roots are
transplanted, as, until that change has
taken place, the process of growth in the
annual renewal of the bulb continues in
progress, and the growth of this new bulb

* See Physiological Sketches, &c., page 136.

is checked by any injury which the leaves or the old bulb may sustain; nevertheless, as it is frequently expedient to remove bulbous plants while their leaves are green, and, even during the time at which they are in flower, this may be safely effected, if done with proper precaution, and also the root may be preserved in .a healthy state, although it will certainly be weakened.　All bulbs, if transplanted while their leaves are in vigour, should be removed with as much soil as will adhere to the bulbs, and great care must be taken not to cut or bruise the root, or the root-fibres.　When transplanted their leaves should be carefully tied to a stick, and suffered to remain until they naturally fall from the plant; if bulbous plants, during their state of vigorous foli-age, are sent to a distance, they should

have the same attention given them, and
the soil should be closely pressed round
the bulbs, and their leaves nicely tied together, and the whole wrapped in sheet
lead, which, by keeping them from the
air, will prevent the evaporation of their
juices, and preserve them for a week or
ten days nearly as well as if they were
placed in soil for that period. As the
leaves of the common hardy kinds of
bulbs give an unneat appearance to gardens, it is a general practice to cut them
off soon after their time of flowering is
over, and if this practice is pursued with
bulbs which have not been planted more
than one or two years, it will weaken
them so much as to prevent their flowering vigorously, and probably destroy the
plant; but when the ordinary kinds of
narcissus, crocuses, and snow-drops, have

continued long in the ground, and are in large patches, their leaves may be cut off when about half decayed, without materially injuring the appearance of the bloom of the ensuing year. The leaves of the more delicate kinds of bulbs must be tied to thin sticks, and the want of neatness occasioned by their withered appearance, borne with, as cutting off the leaves of jonquils, dog's tooth, violet, scillas, hyacinths, &c., would be certain destruction to their roots; and, if the stem of the crown-imperial is not allowed to decay on the bulb from whence it sprang, that bulb will rarely produce flowers. The same theory applies to herbaceous plants, but, as from some particular circumstances, too long to be detailed in this short work, they do not apparently receive equal injury with the

E

bulbous tribe by being deprived of their
leaves, it is not necessary to treat further
on the subject than to suggest to the intelligent Florist carefully to preserve the
foliage of any delicate herbaceous plant
until it spontaneously decays.

M. E. J.

Somersal Hall.

N. B. The generic, specific, and English
names, are given after those of Mr.
Donn's catalogue, that useful publication
being in the hands of most Florists.

THE END.

W. Flint, Printer, Old Bailey, London.

Printed in the United States
By Bookmasters